Block and Graft Copolymers

SAGAMORE ARMY MATERIALS RESEARCH
CONFERENCE PROCEEDINGS
Published by Syracuse University Press

Fundamentals of Deformation Processing
Walter A. Backofen et al., eds.
(9th Proceeding)

Fatigue—An Interdisciplinary Approach
John J. Burke, Norman L. Reed, and Volker Weiss, eds.
(10th Proceeding)

Strengthening Mechanisms—Metals and Ceramics
John J. Burke, Norman L. Reed, and Volker Weiss, eds.
(12th Proceeding)

Surfaces and Interfaces I
Chemical and Physical Characteristics
John J. Burke, Norman L. Reed, and Volker Weiss, eds.
(13th Proceeding)

Surfaces and Interfaces II
Physical and Mechanical Properties
John J. Burke, Norman L. Reed, and Volker Weiss, eds.
(14th Proceeding)

Ultrafine-Grain Ceramics
John J. Burke, Norman L. Reed, and Volker Weiss, eds.
(15th Proceeding)

Ultrafine-Grain Metals
John J. Burke and Volker Weiss, eds.
(16th Proceeding)

Shock Waves and the Mechanical Properties of Solids
John J. Burke and Volker Weiss, eds.
(17th Proceeding)

Powder Metallurgy for High-Performance Applications
John J. Burke and Volker Weiss, eds.
(18th Proceeding)

Block and Graft Copolymers

EDITORS

JOHN J. BURKE

*Planning Director, Army Materials and
Mechanics Research Center*

VOLKER WEISS

Professor, Syracuse University

Proceedings of the 19th Sagamore Army Materials
Research Conference. Held at Sagamore Conference
Center, Raquette Lake, New York, September 5-8,
1972. Sponsored by Army Materials and Mechanics
Research Center, Watertown, Massachusetts, in coop-
eration with Syracuse University. Organized and di-
rected by Army Materials and Mechanics Research
Center in cooperation with Syracuse University.

SYRACUSE UNIVERSITY PRESS 1973

Library of Congress Cataloging in Publication Data

Sagamore Army Materials Research Conference, 19th,
Raquette Lake, N. Y., 1972.
Block and graft copolymers.

(Sagamore Army Materials Research Conference.
Proceedings, 19)
"Sponsored by Army Materials and Mechanics Research
Center . . . in cooperation with Syracuse University."
Includes bibliographical references.
1. Block copolymers—Congresses. 2. Graft
copolymers—Congresses. I. Burke, John J., ed.
II. Weiss, Volker, 1930–   ed. III. United States.
Army Materials and Mechanics Research Center.

Library of Congress Cataloging in Publication Data

IV. Syracuse University. V. Title. VI. Series.
UF526.3.S3 no. 19 [QD382.B5]   623'.028s [547'.84]
ISBN 0-8156-5039-6   73-12903

Printed in the United States of America
Composed and printed by Science Press, Ephrata, Pa.
Bound by Vail-Ballou Press, Inc., Binghamton, N.Y.

To ARTHUR VICTOR TOBOLSKY—friend, teacher, and scientist nonpareil, whose untimely death at the Sagamore Conference in September, 1972, shocked and saddened the worldwide scientific community—this volume is respectfully dedicated.

We cannot hope to do justice to Arthur's lifetime of accomplishment in this brief space, but a review of some highlights of his many-faceted career may help to illustrate the caliber of the man.

As a student: highest honors in physics, chemistry, and mathematics from Columbia; the Jacobus Fellowship at Princeton as the student who, in the judgment of the faculty, showed "the highest scholarly excellence in his graduate work"; Ph.D. from Princeton in physics and physical chemistry at age twenty-five.

As a teacher: on the faculty at Princeton, appointed Associate Professor of Chemistry in 1953; named full professor in 1960; held the Russell Wellman Moore Professorship since 1965.

As an author: his published papers, primarily in the field of polymer science, numbered more than 260; authored or co-authored five books in this or related fields; served on the editorial boards of *American Scientist,* the *Journal of Polymer Chemistry;* and the *Journal of Applied Physics.*

As a professional: served as Chairman of the Division of High Polymers of the American Chemical Society, and as an officer in various other societies; received such awards as the Bingham Medal of the Society of Rheology, the Ford Prize of the American Physical Society, the Witco Award of the American Chemical Society, and the International Award of the Society of Plastics Engineers.

As a consultant: was constantly sought out by the chemical industry in the fields of polymer, physical, and physical organic chemistry, pharmaceuticals, and rocketry.

As a man: a true pioneer in polymers, Arthur Tobolsky contributed to nearly every significant research development since entering this field; but this was not his greatest accomplishment, nor the most valuable part of his legacy. Arthur Tobolsky was first and foremost an extraordinary human being. His scientific work and his writings will live after him. But even more important, the spark of inspiration kindled by his example of complete dedication and fierce desire for excellence will continue to burn brightly, lighting the way for those future scientific leaders who were fortunate enough to have worked with or studied under this uniquely gifted man.

The world is better for having known him.

Herman Mark and Michael Szwarc

Sagamore Conference Committee

Chairman
John J. Burke, Army Materials and Mechanics
Research Center

Program Director
Volker Weiss, Syracuse University

Conference Coordinator
Edward J. Lemay, Army Materials and Mechanics
Research Center

Program Coordinator
Kenneth J. Smith, State University of New York
College of Environmental Science and Forestry

Program Committee
Dr. John J. Burke, Army Materials and Mechanics
Research Center
Dr. Norman G. Gaylord, Gaylord Research Institute, Inc.
Dr. Nathaniel B. Schneider, Army Materials
and Mechanics Research Center
Dr. Kenneth J. Smith, State University of New York
College of Environmental Science and Forestry
Prof. Richard S. Stein, University of Massachusetts
Prof. Michael Szwarc, State University of New York
College of Environmental Science and Forestry
Prof. Arthur V. Tobolsky, Princeton University
Prof. Volker Weiss, Syracuse University

Arrangements at
SAGAMORE CONFERENCE CENTER
John Lathrop, Syracuse University

Contents

SESSION IV
PHYSICAL PROPERTIES
Turner Alfrey, *Moderator*

SESSION V
MECHANICAL AND PHYSICAL PROPERTIES
Arthur V. Tobolsky, *Moderator*

SESSION VI
SPECIAL APPLICATIONS
Vivian Stannett, *Moderator*

Foreword

The Army Materials and Mechanics Research Center has conducted the Sagamore Army Materials Research Conferences in cooperation with the Metallurgical Research Laboratories of the Department of Chemical Engineering and Metallurgy of Syracuse University since 1954. The purpose of the conferences has been to gather together scientists and engineers from academic institutions, industry, and government who are uniquely qualified to explore in depth a subject of importance to the Army, the Department of Defense, and the scientific community.

The principles of copolymerization, having led to new materials already, hold promise for a new era in the development of polymers, possessing unique and unusual properties. Full utilization of such principles, although still in the future, will enable man, for the first time, to prepare polymeric materials having specified characteristics and will endow him with a great capability in the realm of "tailor-made" substances.

This volume, *Block and Graft Copolymers*, addresses the broad areas of synthesis, structure and morphology, physical properties, mechanical properties, and special applications.

The dedicated assistance of Mr. Edward J. Lemay of the Army Materials and Mechanics Research Center throughout all stages of the conference planning and, finally, the publication of the Sagamore Conference proceedings is deeply appreciated. The support of the Technical Reports Office under the supervision of Mrs. A. V. Gallagher, and the Technical Information Office under the supervision of Miss M. M. Murphy of the Army Materials and Mechanics Research Center in preparing the final manuscript is acknowledged.

The continued active interest and support of these conferences by Dr. Alvin E. Gorum, Director, LTC Robert B. Henry, Commander/Deputy Director, and Mr. Edward N. Hegge, Associate Director of the Army Materials and Mechanics Research Center, is appreciated.

Sagamore Conference Center The Editors
Raquette Lake, New York
September 1972

SESSION I

SYNTHESIS AND PROPERTIES

MODERATOR: KENNETH J. SMITH
Chemistry Department
College of Forestry
Syracuse, New York

1. Introduction to the Synthesis and Properties of Block and Graft Copolymers

MICHAEL SZWARC

SUNY College of Environmental Science and Forestry
Syracuse, New York

ABSTRACT

A review of the structural features of block and graft polymers leads to nomenclature describing this class of compounds. This is elaborated. Various synthetic routes leading to the preparation of such polymers are described, and their merits and shortcomings critically examined. Special emphasis is placed on the methods of preparation of graft and block polymers by the living polymer technique. The properties of graft and block polymers are compared with those of the respective polyblends. Various examples are used to illustrate the basic principles emerging from the discussion. The morphology of solid block polymers is briefly surveyed. Finally, the practical application of graft and block polymers is considered, and their future role in modern technology contemplated.

It is my intention to discuss briefly the concept of block and graft polymers, to review the principal methods of their synthesis, and to explain how and why their properties differ from those of ordinary polymers and random copolymers, and their blends.

Arrays of covalently bonded units derived from one or more monomers form molecules of a polymer or copolymer. On a semimicroscopic scale such arrays are homogeneous; i.e., allowing for some fluctuations, the composition and structure of sufficiently long segments, but not too long, are constant and independent of their position along the chain. This homogeneity of ordinary polymers and random copolymers contrasts with the heterogeneity of those macromolecules known as block or graft polymers. The latter involve two or more long segments, and although each segment is homogeneous, its composition or structure differs from that of the adjacent segments. The linear arrays of such segments are referred to as block polymers, and branched structures involving a main chain with side chains of different composition attached to it

1

are known as graft polymers. The semimicroscopic heterogeneity of block and graft polymers endows them with some new and often technologically valuable properties not exhibited by ordinary homopolymers and copolymers, and their blends.

Most block polymers are composed of segments produced by polymerization of two or more different monomers; i.e., macromolecules such as A ... A · B ... B may be formed by consecutive polymerization of A and B monomers. It is also possible to produce block polymers by polymerizing a single monomer. For example, many stereospecific coordination catalysts (Ziegler–Natta catalysts) convert vinyl monomers into isotactic polymers such as

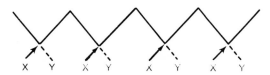

A "mistake" in the polymerization process leads then to stereoblocks

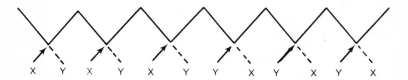

In this type of block polymer the individual segments have identical composition but different *relative* structure, although the *absolute* structure is identical. Such a difference in the *relative* structure of stereoblocks should be contrasted with the genuinely different structure encountered in block polymers formed from the two enantiomorphs of an optically active monomer, e.g., a poly(amino acid) resulting from a condensation of the d and then l alanines, i.e., $A_d \ldots A_d \cdot A_l \cdot A_l \ldots A_l$.

Alternatively, the same monomer may be incorporated into polymer chains in two or more different ways; e.g., butadiene may be linked in 1,4-*cis* or 1,4-*trans* or 1,2 fashion, and thus block polymers composed of segments all 1,4-*trans*- followed by segments of all 1,2- units may be formed from a single monomer. Similarly, acrolein may be polymerized through its C=C or C=O bonds, and, again, this may lead to isomeric blocks in a polyacrolein.

A few words should be devoted here to the nomenclature of block and graft polymers. A concise notation useful in the description of various possible structures of block polymers involves symbols such as $\{A_n\}$ or $\{A_{\bar{n}}\}$ denoting a sequence of n units, or on the average $\bar{n}$ units of A. A block polymer may be composed of two or more blocks (diblock, triblock, etc.), and

$$\{A_n\}\{B_m\} \quad \text{or} \quad \{A_{\bar{n}}\}\{B_{\bar{m}}\}$$

thus denotes a diblock polymer composed of n (or $\bar{n}$) units of A followed by m (or $\bar{m}$) units of B. Similarly,

$$\{A_n\}\{B_m\}\{C_k\} \quad \text{or} \quad \{A_n\}\{B_m\}\{A_n\}$$

denote terblock polymers, etc.

The overall composition and the total length of a macromolecule are not sufficient to uniquely describe a block polymer. For example, the two triblock polymers

$$\{A_n\}\{B_{2m}\}\{A_n\} \quad \text{and} \quad \{B_m\}\{A_{2n}\}\{B_m\}$$

have identical length and composition, but because of the different distribution of monomers along the chain, they represent two distinct macromolecules, often entirely different in their properties.

The individual blocks need not be formed from homopolymers only. Block polymers involving copolymeric blocks are also known, and $\{A\}\{A,B\}$ thus denotes macromolecules composed of a sequence of A units followed by a sequence of randomly distributed A and B units. Closer description of such a block polymer requires not only the specification of the size of the individual blocks but also the composition of the AB copolymer.

Finally, tapered structures deserve some attention. Some preparations yield polymers possessing a block composed of pure A at one end and pure B at the other. These may be linked through a copolymeric block, the composition of which gradually varies from that of pure A to that of pure B. The notation $\{A \longrightarrow B\}$ is proposed for such a "graded" segment, and the triblock polymer described here is thus denoted by $\{A\}\{A \longrightarrow B\}\{B\}$.

The nomenclature of graft polymers is less descriptive than that of block polymers. It is customary to describe them as poly-A-g-poly-B, where poly-A forms the main chain, and poly-B's the branches (grafts). In a more detailed description the average length of the backbone chain (poly-A) and of the branches (poly-B's) are given, as well as the average composition of the macromolecule, i.e., the mole ratio A/B. Such a specification then determines the average number of branches attached to the average backbone chain, leaving unspecified, however, their mode of distribution along the main chain; e.g., are the branches more or less uniformly distributed along the chain or crowded in some sections?

The distribution of branches may profoundly affect the properties of the resulting material, but as far as I know, this problem has never been tackled experimentally. Undoubtedly, a graft polymer such as

$$-\{A_n\}-X-\{A_n\}-X-\{A_n\}-X-, \text{ etc.,}$$
$$\quad\quad | \quad\quad\quad\quad | \quad\quad\quad\quad |$$
$$\{B_m\} \quad\quad \{B_m\} \quad\quad \{B_m\}$$

behaves differently from the star-shaped polymer

$$\{A_{jn}\}-X\!\!-\!\!-\!\!X-\ldots X-\{A_{jn}\},$$
$$\;\;\;\;\;\;\;|\;\;\;\;\;|\;\;\;\;\;\;\;|$$
$$\{B_m\}\;\{B_m\}\;\;\;\{B_m\}$$

especially when $n \gg m$. The aggregation of $\{B_m\}$'s segments would be hindered in the former case but facilitated in the latter. It has been also shown that a few long branches affect the properties of the resulting material differently from many short branches, although the composition of the resulting polymers remains unchanged. It is therefore unfortunate that most of the workers interested in this field, while occasionally extending their inquiry to the determination of backbone length, are satisfied to know only the average composition of the resulting polymer.

Living Polymers

Although the concepts of block and graft polymers were recognized long ago and various techniques leading to their formation were described and explored, the real impetus to study these types of macromolecules was provided by the discovery of living polymers [1,2].

Addition polymerization yields long-chain macromolecules by consecutive linkage of monomeric units to a growing center. Thus,

$$A \cdot A \ldots A \cdot A^* + A \longrightarrow A \cdot A \ldots A \cdot A \cdot A^*.$$
$$n\text{-mer} \hspace{4cm} n + 1\text{-mer}$$

Usually, the growth process, referred to as the propagation of polymerization, is terminated by some reaction that converts the active, "living" unit $-A^*$ into an inactive group $-AX$ or $-AA-$, thus yielding a dead polymer—a macromolecule unable to spontaneously resume growth.

In 1956 we published two notes [1,2] describing the anionic polymerization of some typical monomers that proceed without any termination or chain transfer. The resulting polymers cease to grow when the monomer, initially present in the system, is consumed, although they still retain the ability to resume their growth whenever additional monomer is supplied. (To be more exact, the growth ceases when the concentration of the monomer reaches its limiting "equilibrium" value. Detailed discussion of this problem is presented elsewhere [3].) I therefore proposed the term "living polymers" for such macromolecules to distinguish them from conventional dead polymers formed through some termination or chain-transfer process.

In the same two papers we demonstrated that the growth of a living polymer may be continued through the addition of another monomer; i.e., a living polymer $A \cdot A \ldots A \cdot A^*$ may be "fed" with a suitable monomer B, and then a block polymer $A \cdot A \ldots A \cdot A \cdot B \cdot B \ldots B \cdot B^*$ is formed. Moreover, if the active group $—B^*$ formed after exhaustion of the monomer B are able to add monomer A, chain growth may be continued by supplying monomer A to the system. Thus, a triblock polymer $A \cdot A \ldots A \cdot A \cdot B \cdot B \ldots B \cdot B \cdot A \cdot A \ldots A \cdot A^*$ could be formed. Alternatively, a suitable monomer C may be added to the solution of the living polymer $A \cdot A \ldots A \cdot A \cdot B \cdot B \ldots B \cdot B^*$, and then a triblock polymer

$$A \cdot A \ldots A \cdot A \cdot B \cdot B \ldots B \cdot B \cdot C \cdot C \ldots C \cdot C^*$$

is produced.

We also pointed out [1,2] that it is feasible, e.g., through electron-transfer initiation, to produce living polymers endowned with two growing ends, viz.,

$$*A \cdot A \ldots A \cdot A^*,$$

and that such a polymer fed with a proper monomer B yields directly a triblock polymer,

$$*B \cdot B \ldots B \cdot AA \ldots A \cdot BB \ldots B^*.$$

Why did the discovery of living polymers and the technique of preparation of various types of block polymers discussed above lead to the rapid and fruitful development of the whole field? The obvious reason, stressed in my papers and numerous talks, is the great power of this technique to control the architecture of the resulting polymer and to produce blocks of a desired and well-determined size according to the wishes of the experimenter. The techniques previously employed, which I discuss later, often led to mixtures of homopolymers and block polymers, and even those techniques that produce pure block polymers free of homopolymers do not allow us to control rigorously the size and composition of the resulting macromolecules. Consequently, in the past the investigator was faced with ill-defined materials and was prevented from exercising systematic variation of crucial factors such as the composition or size of block polymers, the distribution of monomers along the chain, etc. It is not surprising, therefore, that research in this field languished before 1956.

The control of polymers formed by the living polymer technique is indeed remarkable. As was pointed out in our early papers, the number average degree of polymerization of living polymers is easily controlled by the supply of monomer to the reactor, viz., $\overline{DP} = M/I$, where M is the number of moles of the monomer supplied, and I is the number of moles of initiator present at the onset of polymerization. (For electron-transfer initiators, $\overline{DP} = 2M/I$.) The monomer may be introduced rapidly at the onset of the reaction, or it may be gradually supplied as the reaction proceeds.

It was shown by Flory [4] that a termination-free polymerization yields polymers of nearly uniform size, i.e., with a Poisson distribution of molecular sizes, provided the initiation is much faster than the propagation. Because initiation of living polymers often is rapid (see, e.g., [5-7] for the various means of accelerating some inherently slow initiations), it is not too difficult to obtain living polymers having virtually uniform and predetermined sizes. Our first attempts to demonstrate the validity of this prediction met with moderate success [8], but eventually the techniques were improved and successful results were reported by numerous investigators (see, e.g., [9-12]). Thus, one can not only predetermine the average size of each individual block in a block polymer but also produce these molecules with blocks of uniform size.

The distribution of monomers in block polymers may be varied at will; e.g., consider the following experiments. Let two reservoirs containing, say, a mole of monomer A and of monomer B, respectively, be attached to a reactor in which a living polymer can be formed. The polymerization may be performed by adding slowly all the A and, thereafter, all the B. Such a procedure yields diblock polymers of 50:50 composition, viz., A . . . A · B . . . B, their degree of polymerization being determined by the amount of initiator present at the onset of the reaction. However, this method of monomer addition is not unique. For example, we may add one-half mole of A, then one-half mole of B, thereafter the remaining one-half mole of A, and finally the residual one-half mole of B. Such a procedure should lead to different block polymers, namely,

$$A \ldots A \cdot B \ldots B \cdot A \ldots A \cdot B \ldots B.$$

The new macromolecules should have the same composition and molecular weight as the previous one but a different monomer distribution along the chain. The generalization of such a procedure is obvious; by regulating the supply of the monomers to the reactor, the distribution of A and B in the polymer may be varied at will, keeping constant, nevertheless, the overall composition and total molecular weight of the product.

Studies of the effect of distribution on the properties of block polymers have therefore become feasible. Such studies were carried out by my former students Levy, and Schlick [13]. They prepared block polymers of styrene and isoprene; they maintained the 1:1 mole ratio of the monomers in the polymers but varied the distribution along the chain. Two series of block polymers were investigated, one having $DP \sim 100$, the other of about 1,000. In each series, the monomers were distributed among 3, 5, 7, and 9 blocks; furthermore, "copolymers" were also prepared by a slow, dropwise addition of the mixture of both monomers to the reactor. It was shown that the solubility and viscosity of these polymers depend on the monomer distribution; i.e., for a constant composition and molecular weight, the distribution of monomers along the chain does affect the properties of the resulting material. Similar results independently obtained by Korotkov et al. were reported [14].

The most spectacular example revealing the profound effect of monomer dis-

tribution on the properties of the polymer produced was provided b
thermoplastic rubbers [15]. These materials, developed by Milk
another of my former students, generated considerable interest i
circles because they demonstrated the technological value of block polymers,
and thus they greatly added to the vigor of research in this field. Thermoplastic
elastomers are triblock polymers—{polystyrene}{polybutadiene or isoprene} ·
{polystyrene}; in short, {S}{B}{S} or {S}{I}{S}—produced by anionic polymerization proceeding in hydrocarbon solvents and involving Li^+ as the
counterion.

The thermoplastic rubbers have considerable tensile strength at ambient
temperatures but become fluid and pliable at elevated temperatures (around
150°C). On the other hand, diblock polymers {S}{B} or {S}{I}, or even triblock polymers containing polystyrene as their middle block, i.e., {B}{S}{B} or
{I}{S}{I}, show no strength and are useless from a technological point of view.
In order to provide the material with the cross-linking necessary to give it tensile
strength, it is essential to terminate *both* ends of the elastomeric blocks by the
glassy type of polystyrene blocks, which act as anchors.

The easy preparation of block polymers through the utilization of anionically-
growing living polymers encouraged investigators to synthesize numerous block
polymers by this route. At present at least 50 different pairs of monomers have
been combined into {A}{B}- and {A}{B}{A}-type block polymers, and many
of them have found practical applications. The variety of materials reported
in the literature includes pairs of hydrocarbons such as styrene–butadiene,
styrene–isoprene, styrene–α-methylstyrene, styrene–dihydronaphthalene, butadiene–isoprene, etc.; hydrocarbons combined with polar monomers, e.g., styrene–
methyl methacrylate (or other acrylates), styrene–acrylonitrile, styrene–
ethylene oxide, styrene–formaldehyde, styrene–propylene sulfide, etc.; combinations of two polar monomers, e.g., acrylonitrile-*n*-butylisocyanate, ethylene
sulfide–propylene sulfide, ethylene oxide–propylene oxide; and so on. However,
not every monomer capable of forming living polymers reacts with all known
living polymers; e.g., styrene cannot be added to living poly(methyl methacrylate), although methyl methacrylate can be added to living polystyrene [17].

Living polymers offer other advantages to synthetic chemists. The existence
of active ends permits the conversion of a nonfunctional polymer into a macromolecule endowed with a functional end-group. This opportunity, stressed in
my early papers [19], was first explored by Richards and Szwarc [20] who
prepared samples of polystyrenes identical in every respect, but endowed
with different end-groups, viz., terminated by $-CH_2(Ph)$, carboxylic or hydroxylic groups. Since then many polymers with a variety of end-groups have
been prepared and reported in the literature. Such polymers may be used
as building blocks for the preparation of block polymers by the condensation
technique, an approach particularly interesting when applied to living poly-

mers having both ends active. These may be converted into bifunctional polymers and used in the preparation of polyblock polymers, e.g., $nX\{A\}X + nY\{B\}Y \longrightarrow -\{A\}Z\{B\}Z\{A\}-$, etc. Alternatively, by utilizing bifunctional coupling agents, two living polymers may be linked into a block polymer. The latter technique was used by Milkovich [15,16] in his preparation of a thermoplastic rubber. He linked diblock living polymers $\{S\}\{B\}^-$, Li^+ with reagents such as CH_2Br_2 into $\{S\}\{B\}CH_2\{B\}\{S\}$. Coupling of living polymers having both ends active leads again to polyblock polymers or to some polymers organized in an unusual way. In the hands of Richards and his colleagues [21], similar methods of coupling led to a variety of most interesting products, as well as to new types of polymers; e.g., coupling of dimeric α-methylstyrene dianions with iodine results in a head-to-head tail-to-tail poly-α-methylstyrene,

$$-CH_2 \cdot C(Ph)(CH_3) \cdot C(Ph)(CH_3) \cdot CH_2CH_2C(Ph)(CH_3)C(Ph)(CH_3)CH_2CH_2-,$$

a material different from the conventional head-to-tail poly-α-methylstyrene.

Reactions of living polymers with suitable groups located on other polymers result in grafting. For example, an ordinary random copolymer of styrene and chloromethylated styrene may be reacted with living butadiene to yield a graft polymer. If the living polymer is uniform in size, all the branches of the resulting graft polymer are of uniform length. Such macromolecules are referred to as comb-shaped graft polymers.

Some years ago I outlined a technique that should lead to uniform distribution of branches along the main chain [19]. This approach is illustrated by the following hypothetical synthesis. Bifunctional living polymers of uniform size, say $\{A_n\}$ terminated on both ends by carbanions, are reacted with

$$\begin{array}{c} BrCH_2Ph-C-CH_2-PhCH_2Br. \\ \parallel \\ CH_2 \end{array}$$

This leads to coupling and yields the polymer

$$\left[\begin{array}{c} (A_n)-CH_2Ph-CH_2-C-CH_2PhCH_2^- \\ \parallel \\ CH_2 \end{array} \right]_n,$$

which could be oxidized to

$$-(A_n)-X\underset{\parallel}{C}X-(A_n)-X\underset{\parallel}{C}X-(A_n)-,$$
$$OO$$

X denoting $-CH_2PhCH_2-$ groups. The resulting polymer therefore contains carbonyl groups uniformly distributed along its chain. The latter could be reacted with living polymers of uniform size, say poly-B, each possessing only one active carbanion end-group. Then a branched polymer,

$$-(A_n)-X\underset{\underset{(B_m)}{|}}{C}(OH)X-(A_n)-X\underset{\underset{(B_m)}{|}}{C}OHX-(A_n)-,$$

could be produced having branches of uniform size regularly distributed along the chain.

The above synthesis provides an illustration of the method leading to the uniform comb-shaped polymers. Obviously, other, more practical variants of this technique may be developed in the future. To my knowledge such syntheses have not been reported yet in the literature.

Finally, another class of polymers exhibiting a semimicroscopic heterogeneity should be mentioned here. These are the so-called star-shaped polymers in which several polymer arms radiate from a common center. The ordinary homogeneous star-shaped polymers are represented by the formula

$$G(A_n)_j \quad \text{or} \quad G(A_n)_k(A_m)_j, \text{etc.,}$$

where G denotes some polyfunctional group to which j or j and k polymer arms are attached, their degree of polymerization being n or m, respectively.

Living polymers provide the best route to star-shaped polymers. Two techniques can be employed. Either a polyfunctional initiator is used—e.g., methyl acetylene reacted with butyllithium yields $Li_2C{=}C{=}CLi_2$, which may initiate polymerization of some polar monomer and produce a four-armed star-shaped polymer [22]—or, alternatively, a polyfunctional terminator is used to couple linear living polymers having only one growing end-group to a star-shaped polymer. The latter method was employed by several investigators who used terminating agents such as $SiCl_4$ [23], trichloromethyl benzene [24], the cyclic trimer of phosphonitrilic chloride [25], etc., and in spite of some difficulties, the desired polymer structures have been obtained.

A modification of the latter technique may be utilized in the preparation of heterogeneous star-shaped polymers, i.e.,

$$G(A_n)_j(B_m)_k.$$

I outlined the general approach to such a synthesis elsewhere [19], and the first preparation of heterogeneous star-shaped polymers was reported by Stannett and his associates [26].

Although all the heterogeneous polymers discussed above were formed by anionic polymerization, the application of living polymers for synthesis of block polymers is not restricted to the conventional anionic mode of propagation. In fact, any system in which propagation proceeds without termination or chain transfer may be utilized in the preparation of block polymers. One may be even less restrictive and be satisfied with a system in which termination and chain transfer are infrequent within the time needed for completion of the synthesis.

Polymerization of Leuchs' anhydrides exemplifies an unconventional anionic propagation [18] that may lead to block polymers containing polypeptides. There are cationically-growing systems in which termination and chain transfer are avoided, and these are suitable for the preparation of block polymers. The cationic polymerization of tetrahydrofuran and other cyclic ethers serves as an example [27]. The lifetime of polymers growing on some heterogeneous co-ordination catalysts is sufficiently long to permit the preparation of block polymers, and such preparations have also been reported in the literature [28,29].

In radical polymerization, the termination is caused by bimolecular collisions between two growing radicals, which results in coupling or disproportion-ation. Thus, termination cannot be prevented as long as the system is fluid and the polymer soluble. However, in some systems the polymer precipitates as it is formed, and then the growing radical is trapped in the polymer. This phenomenon prevents termination but not propagation, because the monomer may diffuse into the particle and "feed" the trapped radical. Such systems are therefore suitable for the preparation of block polymers, and some examples of these phenomena are found in the literature [30].

A somewhat similar situation is encountered in rigid matrices in which mono-mer is dissolved. Trapped radicals may be formed, e.g., by γ-irradiation, and these become essentially immobile after a few monomer molecules have been added to the growing center. Thus, the termination, but not the feeding, may be prevented, and such a system therefore could be utilized in the preparation of block polymers. Alkaline ice in which some hydroperoxide is dissolved may provide a particularly convenient system for such a synthesis.

Other Methods of Synthesis of Block and Graft Polymers

Two basic strategems are followed in the preparation of block polymers. Either one block is produced first and then the polymerization of the other monomer is initiated by an active or activated end-group of the previously pre-pared block, or both blocks are prepared and then linked together by some suitable reaction.

The synthesis of block polymers with the aid of living polymers exemplifies the first approach. However, this route also leads to block polymers when cer-tain dead polymers are used, provided their end-groups can be activated. For example, radical polymerization in the presence of CCl_3Br (a chain-transfer agent) yields polymers such as

$$Br \cdot AA \ldots AA \cdot CCl_3.$$

Such polymers may be photolyzed in the presence of another monomer, say B, and the resulting macroradical $CCl_3 \cdot A \cdot A \ldots A \cdot A \cdot$ can initiate polymerization of B. Thus, the reaction yields a block polymer $CCl_3 AA \ldots A \cdot BB \ldots B$. Alternatively, the initiator used for the preparation of the first polymer may involve a photolyzable or thermally labile group that becomes attached to the resulting polymer. Decomposition of such a polymer by light or heat in the presence of a suitable monomer B yields a block polymer $\{A\}\{B\}$. However, such techniques generate two radicals—a macroradical that is converted to a block polymer, and a small fragment producing a homo–B–polymer that contaminates the block polymer.

This drawback of the above method may be avoided by using redox systems for the generation of radicals. For example, a dead polymer with a terminal $-CH_2OH$ yields a macroradical with Ce^{4+} salts, i.e.,

$$AA \ldots A \cdot CH_2OH + Ce^{4+} \longrightarrow AA \ldots ACH_2O \cdot + H^+ + Ce^{3+}$$

or

$$AA \ldots A\overset{\cdot}{C}HOH + H^+ + Ce^{3+}.$$

Alternatively, a polymer possessing a hydroperoxy group may generate a macroradical by reacting with Fe^{2+} salt:

$$AA \ldots AROOH + Fe^{2+} \longrightarrow AA \ldots ARO \cdot + OH^- + Fe^{3+}.$$

Similarly, photolysis of a copolymer of A with small fractions of vinyl methyl ketone as the comonomer, e.g.,

$$-AA \ldots A \cdot CH_2{-}\underset{\underset{\displaystyle COCH_3}{|}}{CH} \cdot AA \ldots ACH_2{-}\underset{\underset{\displaystyle COCH_3}{|}}{CH} \cdot AA \ldots A{-},$$

performed in the presence of a suitable monomer B yields a graft polymer, poly-A-g-poly-B.

The redox systems are particularly convenient for the preparation of graft polymers. An excellent technique based on direct peroxidation of various polymers was developed by Mesrobian and Mark [31]. Grafting with the aid of ceric salts of polymers containing hydroxylic groups, e.g., cellulose or copolymers involving $-CH_2 \cdot CH(OH)-$ segments, produces graft polymers free of homopolymers [32].

The halogen–containing polymers, e.g., polyvinylchloride or its copolymers, may be activated by cationic initiators. This method of grafting (or blocking) turned out to be particularly valuable in the hands of Kennedy [33] who prepared, e.g., poly(vinyl chloride)-g-polyisobutene by coordinating the $-CHCl-$ group with Et_2AlCl:

$$-CHCl- + Et_2AlCl \longrightarrow -\overset{+}{C}H-.$$
$$Et_2AlCl_2^-$$

Macroradicals generating block polymers may also be formed by rupture of a polymer chain by swelling, mechanical degradation, etc. This method yields complex mixtures of homopolymers and block polymers when the bond fission takes place in the presence of monomers. Much work in this field has been done by Ceresa, whose monograph [34] should be consulted for further details.

Finally, numerous graft polymers were prepared by creating reactive centers on desired polymers through γ-irradiation. This subject is much too broad to be discussed here. An extensive review including numerous references was published by Battaerd and Tregear [35], and some recent developments in this field are summarized by Stannett [36].

The other strategem—the linking of two polymers into a block—has been briefly discussed in the preceding section. There we were concerned with living polymers; however, this technique has been applied extensively to other polymers having suitable end-groups, and an authoritative review of this subject was published by Smets [37].

Properties of Block and Graft Polymers

All the novel and interesting properties of block and graft polymers stem from their semimicroscopic heterogeneity. Each block retains virtually all the characteristics of the respective homopolymers or copolymers, and becuase the individual components are usually incompatible, the blocks of the graft or block polymer try to avoid each other. This kind of segregation is reflected in the behavior of dilute solutions of block and graft polymers; much of the experimental and theoretical work encompassing this subject was done by Benoit and his associates [38].

The most spectacular effects arising from this incompatibility are observed in concentrated solutions, melts, and solid materials. The tendency to segregate the individual components leads to the formation of domains of {A}'s, {B}'s, etc. However, because the incompatible blocks are linked by chemical bonds, the domains have well-defined sizes and shapes determined by the length of the individual blocks, the composition of the solution or melt, temperature, etc. Thus, mesomorphic phases are formed; these were first recognized and investigated by Sadron's group in the course of their pioneering studies of block polymers of styrene and ethylene oxide [39].

The mesomorphic phases may form lamellae, e.g., concentrated solutions of {styrene}{butadiene} diblocks in ethyl benzene, or cylindric or spherical particles containing one polymer dispersed in the continuum involving the other. The regularity of such forms is sometimes remarkable, and for the sake of illustration, electron micrographs of some solid materials composed of block polymers prepared by Galot and Sadron are shown in Figure 1.

Figure 1. Electron micrographs of some solid materials composed of block polymers (prepared by Galot and Sadron).

The structure of mesomorphic phases and, therefore, the properties of the resulting material also depend on the handling of the polymer after its preparation. For example, the properties of a film formed from rubber with grafts of polyethylmethacrylate depend on the nature of solvent from which the film is cast. Thus, rubbery or glassy properties could be imparted to the same material.

The morphologic structures developed in solid block and graft polymers endow these materials with new properties. The polystyrene domains, which act as multiple cross-links, endow thermoplastic rubber with its strength. The domains seem to be preserved even in the melt at 150°C. The high melt viscosity of these materials is accounted for by their presence.

The cylindric domains shown in Figure 1 may act like fiberglass in fiberglass-reinforced plastic, or as wires in steel-reinforced cement. Indeed, this aspect of the problem is still unexplored, and the above materials may prove most valuable from a technological point of view.

Block and graft polymers differ from the respective blends becuase in the latter the segregation is not limited to microscopic domains. In fact, a complete separation into two phases is, thermodynamically, the most stable state of such systems. Block polymers therefore act as emulsifiers for polymeric blends and may stabilize their dispersion at the required level. Again, this may lead to new and valuable materials.

Finally, the heterogeneity of block or graft polymers, amplified by the formation of mesomorphic domains, allows each component of such systems to retain its basic properties. Thus, one can upgrade a material composed of polymer A that has a specific valuable property, call it P, but that lacks some other required property, Q, by forming a block or graft polymer of {A} and {B}, where {B} has the required property Q. Such a procedure may endow the new material with the property Q without affecting its chief performance with respect to the property P. As an example, we may cite membranes designed for desalination prepared by Stannett and his associates [40]. The conventional cellulose acetate membrane, used in reverse osmosis, suffers from compaction under the pressure applied in the process, which in time reduces its permeability. Polystyrene is characterized by its low rate of creep, and the grafting of polystyrene on cellulose acetate thus greatly improves the mechanical properties of the latter, essentially without affecting its permeability.

A word of warning is in order. One should recall the famous conversation of Bernard Shaw with a remarkably beautiful lady known also for her stupidity. "Let us have a child," she said, "and imagine it will inherit my beauty and your wisdom." Bernard Shaw was dubious. "But what happens if the child will be as ugly as I and as stupid as you?" he answered. This is, of course, a real danger. Nevertheless, I hope that the results reported in this meeting, as well as future work, will prove that there are children who inherit the beauty of the mothers and the wisdom of the fathers.

Acknowledgment

The author wishes to thank all his students and co-workers who contributed so much to studies of living polymers, and the National Science Foundation for their generous support of that research.

References

1. Szwarc, M., Levy, M., and Milkovich, R., "Polymerization Initiated by Electron Transfer to Monomer. A New Method of Formation of Block Polymers," *J. Amer. Chem. Soc.,* 78 (1956), 2656.
2. Szwarc, M., "Living Polymers," *Nature (London),* 178 (1956), 1168.
3. Szwarc, M., "Thermodynamics of Polymerization with Special Emphasis on Living Polymers," *Adv. Polym. Sci.,* 4 (1967), 457, and *Carbanions, Living Polymers and Electron Transfer Processes,* New York: Interscience (1968), 104.
4. Flory, P.J., "Molecular Size Distribution in Ethylene Oxide Polymers," *J. Amer. Chem. Soc.,* 62 (1940), 1561.
5. Bywater, S. and Worsfold, D.J., "Anionic Polymerization of Styrene. Effect of Tetrahydrofuran," *Can. J. Chem.,* 40 (1962), 1564.

6. Burnett, G.M. and Young, R.N., "The Polymerization of Substituted Styrenes by Butyl Lithium. I. The Initiation Reaction," *Eur. Polym. J., 2* (1966), 329.

7. Bywater, S. and Worsfold, D.J., "Alkyllithium Anionic Polymerization Initiators in Hydrocarbon Solvents," *J. Organometal. Chem.,* 10 (1967), 1.

8. Waack, R., Rembaum, A., Coombes, J.A., and Szwarc, M., "Molecular Weights of 'Living' Polymers," *J. Amer. Chem. Soc.,* 79 (1957), 2026.

9. Sirianni, A.F., Worsfold, D.J., and Bywater, S., "Anionic Polymerization of α-Methyl Styrene. Part 3. Molecular Weight Determinations on Sharp Distribution Polymers," *Trans. Faraday Soc.,* 55 (1959), 2124.

10. Altavers, A.F., Wyman, D.P., and Allen, V.R., "Synthesis of Low Molecular Weight Polystyrene by Anionic Techniques and Intrinsic Viscosity–Molecular Weight Relations Over a Broad Range in Molecular Weight," *J. Polym. Sci., Part A,* 2 (1964), 4533.

11. Yen, S.P.S., "Synthesis of Virtually Monodisperse Linear and Tetrachain Polymers with Idealized Structures by Use of a Novel Initiator, α-Phenylethylpotassium," *Makromol. Chem.,* 81 (1967), 152.

12. Böhm, L., Barnikol, W.K.R., and Schulz, G.V., "Kinetic Verification of Two Types of Ion Pairs in the Anionic Polymerization of Styrene in Tetrahydropuran," *Makromol. Chem.,* 110 (1967), 222.

13. Schlick, S. and Levy, M., "Block-Polymers of Styrene and Isoprene with Variable Distribution of Monomers Along the Polymeric Chain. Synthesis and Properties," *J. Phys. Chem.,* 64 (1960), 883.

14. Korotkov, A.A., Shibayer, L.A., Pyrkov, L.M., Aldoshin, V.G., and Frenkel, S., "The Synthesis and Investigation of Hybrid Polymers. I. Styrene and Isoprene Block Polymers Formed by Catalytic Polymerization in Solution under the Action of Butyl Lithium," *Vysokomol. Soedin.,* 1 (1960), 157.

15. Holden, G., Bishop, E.T., and Legge, N.R., "Thermoplastic Elastomers," *J. Polym. Sci., Part C,* no. 26 (1969), 37.

16. U.S. Patent No. 3,265,765, August 9, 1966, "Block Polymers of Monovinyl Aromatic Hydrocarbons and Conjugated Dienes," G. Holden and R. Milkovich to the Shell Oil Company, New York, New York.

17. Graham, R.K., Dunkelberger, D.L., and Good, W.E., "Anionic Copolymerization: The Inability of the Poly(methyl Methacrylate) Anion to Initiate the Polymerization of Styrene," *J. Amer. Chem. Soc.,* 82 (1960), 400.

18. Szwarc, M., "The Kinetics and Mechanism of N-carboxy–α-amino acid Anhydride (NCA) Polymerisation to Poly–amino acids," *Adv. Polym. Sci.,* 4 (1965), 1.

19. Szwarc, M., "Recent Advances in Polymer Chemistry," *Adv. Chem. Phys.,* 2 (1959), 147, and "Termination of Anionic Polymerization," *Adv. Polym. Sci.,* 2 (1960), 275.

20. Richards, D.H. and Szwarc, M., "Block Polymers of Ethylene Dioxide and Its Analogues with Styrene," *Trans. Faraday Soc.,* 55 (1959), 1644.

21. Cunliffe, A., Hubbert, W.J., and Richards, D.H., "Synthesis of Copolymers. I. Copolymers of Styrene and of α-Methylstyrene," *Makromol. Chem.,* 157 (1972), 23.

22. West, R., Carney, P.A., and Mineo, I.C., "The Tetralithium Derivative of Propyne and Its Use in Synthesis of Polysilicon Compounds," *J. Amer. Chem. Soc.,* 87 (1965), 3788.

23. Morton, M., Helminiak, T.E., Gadkary, S.D., and Bueche, F., "Preparation and Properties of Monodisperse Branched Polystyrene," *J. Polym. Sci.,* 57 (1962), 471.

24. Orifino, T.A. and Wenger, F., "Dilute Solution Properties of Branched Polymers. Polystyrene Trifunctional Star Molecules," *J. Phys. Chem.,* 67 (1963), 566.

25. Altares, T.A., Wyman, D.P., Allen, V.R., and Myersen, K., "Preparation and Characterization of Some Star- and Comb-Type Branched Polystyrenes," *J. Polym. Sci., Part A,* 3 (1965), 4131.
26. Gervasi, J.A., Gosnell, A.B., Woods, D.K., and Stannett, V., "Reactions of Living Polystyrene with Difunctional Nitriles to Produce Specifically Placed Grafting Sites," *J. Polym. Sci., Part A-1,* 6 (1968), 859.
27. Dreyfuss, P. and Dreyfuss, M.P., "Polytetrahydrofuran," *Adv. Polym. Sci.,* 4 (1967), 528.
28. Natta, G., "Properties of Isostatic, Atactic and Stereoblock Homopolymers, Random and Block Copolymers," *J. Polym. Sci.,* 34 (1959), 531.
    French Patent No. 1,220,573, May 25, 1960, *Copolymèrs Hétérobloc et Procédé pour les Prépare,* G. Natta, E. Grachetti, and I. Pasquon to Montecatini.
29. Bier, G., Lehmann, G., and Lengering, H.J., "Block polymerisate aus Äthylen und Propylen," *Makromol. Chem.,* 45 (1961), 347.
    Bier, G., "Hochmolekulars Olefin-Mischpolymerisate hergestellt unter Verwendung von Ziegler-Misch-katalysatoren," *Angew. Chem.* 73 (1961), 186.
30. U.S. Patent No. 2,666,042, January 12, 1954, *Copolymer Preparation,* K. Nozaki to the Shell Development Company, Emoryville, California.
31. Mesrobian, R.B. and Mark, H., "Grafting with the Aid of Ceric Salts," *Text. J.,* 23 (1963), 294.
32. Mino, G. and Kaizerman, S., "New Method for the Preparation of Graft Copolymers. Polymerization Initiated by Ceric Ion Redox Systems," *J. Polym. Sci.,* 31 (1958), 242.
33. U.S. Patent No. 2,565,878, February 23, 1971, "Process to Prepare a Novel Heat Stabilized Halogen Containing Polymer," J.P. Kennedy to Esso Research and Engineering Company, Linden, New Jersey.
34. Ceresa, R.J., *Block and Graft Copolymers,* London: Butterworths (1962).
35. Battaerd, H.A.J. and Tregear, G.W., *Graft Copolymers,* New York: Interscience (1967).
36. Stannett, V.T., "Some Recent Advances in Pure and Applied Aspects of Graft Copolymers," *J. Macromol. Sci., Chem.,* A4, no. 5 (1970), 1177.
37. Smets, G., "Graft Reactions on Polymers and Comparison of the Various Methods in Use," *Chem. Weekbl.,* 56, no. 12 (1960), 177.
38. Benoit, H., *J. Polym. Sci.,* C4 (1964), 1589.
39. Sadron, C., *Agnew. Chem.,* 75 (1963), 472.
40. Stannett, V., *Polymer Preprints,* 13 (1972), 747.

SESSION II

SYNTHESIS

MODERATOR: JOHN K. STILLE
Department of Chemistry
University of Iowa
Iowa City, Iowa

2. Effects of Polymers on Grafting

NORMAN G. GAYLORD

Gaylord Research Institute, Inc.
New Providence, New Jersey

ABSTRACT

The polymer upon which a graft copolymerization reaction is carried out is generally considered solely as the substrate upon which a reactive site is generated. However, polymer–monomer interactions resulting in salt formation, hydrogen bonding, stereoregular fit, and charge-transfer complexation may lead to changes in the monomer conversion, polymerization mechanism, grafting mechanism, grafting efficiency, and grafted polymer composition, structure, and tacticity.

The polymerization of monomer molecules in solution proceeds at a rate dependent upon the concentrations of monomer and initiator. The rate is dependent upon the temperature to the extent that the number of initiating species, e.g., free radicals, is dependent upon the rate of generation of the species as a function of temperature and to the extent that the propagation and termination reactions are temperature-dependent.

The polymerization, after the initial period, is characterized by a steady state wherein the presence of polymer produced at lower levels of conversion apparently does not influence the rate of polymerization in subsequent periods and, in the case of copolymerization, does not influence the structure of the copolymer produced in these periods. A change in copolymer composition as a function of conversion is due to the change in monomer composition resulting from differing comonomer reactivities, and is not due to the presence of polymer.

At nominal conversions the presence of polymer per se does not influence the molecular weight of the polymer subsequently formed. At high conversions an increase in the viscosity of the reaction medium due to the presence of dissolved polymer results in an increase in the rate and molecular weight of the polymer being formed as the termination reaction becomes diffusion-controlled, and the probability of termination by the collision of polymer chain-ends decreases, while the rate of diffusion of monomer to the chain-ends decreases to a much smaller extent.

19

An increase in the rate of polymerization and in the molecular weight of the polymer is noted during precipitation polymerization, in which the polymer is insoluble in the monomer and/or the reaction medium. The increase is not due to the presence of the polymer per se but rather to the continuation of the polymerization in the precipitated polymer containing trapped reactive chain-ends and monomer in the high-viscosity medium created by the polymer. Thus, the presence of a polymer per se apparently does not change the reactivity of a monomer or of the chain-end derived therefrom in homopolymerization or copolymerization.

Notwithstanding the implied nonparticipation of a polymer during the polymerization of a monomer, active sites may be generated on a polymer as a result of attack by an initiating species or chain transfer by a propagating chain-end. The resultant active site on the polymer then adds monomer molecules to form a branched polymer. When these active sites are generated on a preformed polymer of composition different from that resulting from the polymerization of the monomer that is present in the reaction mixture, a graft copolymer is produced. Nevertheless, the preformed polymer is generally considered solely as the substrate upon which a reactive site is generated, and its only effect on the monomer polymerization may be an increase in the rate of polymerization or the molecular weight of the newly formed polymer due to the increased viscosity, or gel effect. However, recent findings have indicated that monomer–polymer interactions may lead to changes in the polymerization mechanism, grafting mechanism, monomer conversion, grafting efficiency, and grafted polymer composition, structure, and tacticity.

Polymer–Monomer Interaction Through Salt Formation

Spontaneous polymerization of 4-vinylpyridine in aqueous systems occurs in the presence of polymeric acids such as poly(styrenesulfonic acid), poly(acrylic acid), and poly(glutamic acid) [1–4]. The 4-vinylpyridine is adsorbed in equimolar quantities on the polymeric acid substrate, and salts are formed at the sites of interaction and polymerize spontaneously. The monomer molecules in excess of the stoichiometric quantity are unable to fit on the polymeric matrix and, in the absence of activation, do not participate in the polymerization.

The alignment of 4-vinylpyridine molecules on the polymeric substrate is shown in Equation (1) on the following page.

The interaction and polymerization occur even on mixing extremely dilute solutions of the polymeric acid and the monomer because, independent of the average monomer concentration in the solution, the activated monomer molecules are clustered along the chains of the polymeric activator. The rate of

$$(1)$$

polymerization is strongly dependent upon the degree of neutralization of the polyacid, i.e., the degree of filling of the matrix, and reaches a maximum when the matrix is essentially completely neutralized.

If the polyacid matrix is partially neutralized with a nonpolymerizable base such as NaOH before the addition of the vinylpyridine monomer, the matrix is "spoiled" and the rate of polymerization is greatly reduced as compared to the situation with an "unspoiled" matrix. The kinetic data indicate that the vinyl-pyridine molecules are not firmly fixed on the polyacid but have sufficient mobility to migrate along the chain.

The morphology of the reaction product is related to the shape of the matrix in solution. Thus, polymerization of 4-vinylpyridine in the presence of poly-(styrenesulfonic acid) in methanol, where the polyacid matrix is coiled, results in the formation of irregular globular aggregates. However, polymerization in water, where the polyacid matrix is highly ionized and straightened, produces fibrillar aggregates [1].

Although the spontaneous polymerization of 4-vinylpyridine on poly(acrylic acid) occurs at a rapid rate to yield a rigid aggregate containing equivalent quantities of polyacid and polybase [5-7], little or no polymerization occurs when vinylpyridine monomer is added to an aqueous acetic acid solution. It is apparent that the greatly accelerated polymerization rate in the presence of the high-molecular-weight polyacid involves more than a concentration effect.

The initial rate of polymerization of 4-vinylpyridine on poly(acrylic acid) is dependent upon the pH of the reaction medium, which influences the uncoiling of the polyacid coils and their ability to bind cations [6,7]. The rate of poly-merization is reduced by the presence of sodium chloride as the sodium cations compete with the monomer cations for adsorption sites at the active ends of the growing chains [7]. The influence of matrix configuration is clearly shown when the monomer is polymerized in the presence of poly-L-glutamic acid [7].

The rate of polymerization sharply decreases when the pH is changed from 5 to 6, i.e., as the matrix chain changes from a helix to a coil, which has considerably increased spaces between the carboxyl groups.

The products of the spontaneous polymerization of 4-vinylpyridine in the presence of low-molecular-weight and polymeric acids were originally considered to be poly(4-vinylpyridinium) salts, as is actually the case in the polymerization of N-alkyl-4-vinylpyridinium salts, as shown in Equation (2). However, it was subsequently shown that the products are ionenes, as shown in Equation (1).

$$
\begin{array}{ccc}
\text{CH}_2\text{=CH} & & \text{\textasciitilde CH}_2\text{–CH\textasciitilde} \\
\text{pyridine ring} & \longrightarrow & \text{pyridine ring} \\
\text{N+} & & \text{N+} \\
\text{R} \quad \text{X}^- & & \text{R} \quad \text{X}^-
\end{array} \qquad (2)
$$

The effect of developing high localized concentrations of vinylpyridine in the presence of a polymeric acid has been demonstrated in the preparation of semipermeable membranes containing both carboxylic and pyridine groups [8,9].

Acrylic acid is grafted onto polytetrafluoroethylene films by irradiating the latter immersed in aqueous solutions of the monomer with cobalt-60 gamma rays. When the grafted film is immersed in 4-vinylpyridine, spontaneous polymerization occurs, presumably to the ionene structure, shown in Equation (1). Treatment of the acrylic acid–grafted film with an aqueous solution of potassium hydroxide converts the carboxylic groups into potassium salts and prevents the spontaneous polymerization when the film is immersed in 4-vinylpyridine. However, upon irradiation the latter is grafted into the film.

Although the structure of the grafted chains has not been determined, it is probable that the structure is that of polyvinylpyridine, since ionene formation requires protonation of the monomer. Nevertheless, in the presence of the poly(potassium acrylate) branches, 4-vinylpyridine polymerizes considerably

TABLE I

Grafting of 4-Vinylpyridine into Polytetrafluoroethylene (PTFE) Films
Containing Grafted Acrylic Acid (AA)*

AA in PTFE Film (%)	Time for 100% Grafting (min)	Total Dose[†] (rads)
0	2,880	138,000
16	21	1,000
23	14	670
35	<6	<300

*Acrylic acid branches neutralized with KOH.
[†]Does rate, 48 rads/min at 20°C.

faster than in ungrafted polytetrafluoroethylene films (Table I). An extremely small radiation dose leads to very high grafting ratios indicating faster chain propagation, presumably due to the presence of high localized concentrations of the monomer aligned along the polymeric matrix.

Polymer–Monomer Interaction Through Hydrogen Bonding

The polymerization of acrylic acid in aqueous solution using potassium persulfate as initiator at 75°C proceeds fairly slowly. The rate is dramatically increased in the presence of poly(vinylpyrrolidone) (PVP), reaching a maximum when the acrylic acid and the polymer are present in slightly less than equimolar proportions. No change in rate is observed when methylpyrrolidone is substituted for PVP [10].

The intrinsic viscosity of the PVP–PAA complex is determined by the intrinsic viscosity of the PVP. The PVP molecules apparently adsorb the monomeric acid, creating high local concentrations. Radical initiation of polymerization results in a "zipping up" of the acrylic acid molecules. The PAA–PVP complex precipitates at a critical molecular weight of the polyacid. Paper electrophoresis separates the complex into PAA, PVP, and a PVP–PAA graft copolymer [10].

The interaction of PVP and acrylic acid presumably involves hydrogen bonding, as shown in Equation (3):

$$(3)$$

The formation of graft copolymer apparently involves generation of a radical site on the PVP chain by reaction with a radical derived from the initiator or by chain transfer with a propagating PAA chain.

The polymerization of acrylic acid in the presence of PVP in dilute aqueous solution using potassium persulfate as initiator at 65°C, or in bulk (36 weight-percent PVP) using azobisisobutyronitrile as initiator at 60°C yields PAA having the same atactic structure as the polymer obtained by radical polymerization in the absence of PVP [11]. In contrast, methacrylic acid yields normal atactic radical polymer when polymerized in dilute aqueous solution in the presence of PVP, but yields isotactic polymer when polymerized in bulk in the presence of PVP.

The formation of isotactic polymer suggests that the methyl group has a steric effect in the alignment of methacrylic acid molecules along the PVP matrix. This is particularly significant since radical polymerization of vinyl monomers yields atactic polymer at normal temperatures and syndiotactic polymers at low temperatures.

The radical polymerizations of methyl acrylate and methyl methacrylate in the presence of PVP, either in aqueous solution or in bulk, yield polymers having the same tacticities as those obtained in the absence of PVP [11]. The polymerization of vinyl acetate with benzoyl peroxide at 70°C in the presence of PVP similarly yields normal radical polymer [12].

The free-radical polymerization of N-vinyl-2-oxazolidone (NVO) at 60°C in dilute aqueous solution, using potassium persulfate as initiator, in the presence of poly(methacrylic acid) (PMAA) achieves a maximum rate at a 1:1 mole ratio of NVO and PMAA and decreases above 1:1. The rate of polymerization is not affected by the molecular weight of the PMAA, but the molecular weight of the PNVO increases as the molecular weight of the PMAA increases. The polymer complex of PNVO with PMAA precipitates at about 50 percent conversion of NVO [13].

The interaction of NVO and PMAA apparently involves hydrogen bonding, as

$$\text{(4)}$$

shown in Equation (4), since polymers containing a 2-oxazolidone moiety form stable complexes with compounds containing ionizable hydrogen, e.g., carboxylic acids [13].

Since the rate of polymerization was followed by determination of the amount of unreacted NVO and the molecular weight of the PNVO was determined by extracting the latter from a solution of the polymer complex, the presence of graft copolymer was not investigated [13].

The potassium-persulfate-initiated polymerization of acrylic acid in aqueous solution in the presence of poly(ethyleneimine) (PEI) at 30°C is characterized by a greatly increased rate of polymerization as compared to the polymerization in the absence of the PEI. The maximum rate is found at PEI concentrations of about 25 percent of that representing equimolar proportions of PEI and acrylic acid. This may be due to the precipitation of the complex. Electrophoresis separates the complex into PAA and PEI. No graft copolymer is detected [14].

When the photoinitiated polymerization of acrylic acid in the presence of PEI is carried out at 25°C in an acetone–water mixture, using azobisisobutyronitrile as photosensitizer, the system remains homogeneous. The initial rate of polymerization increases with increasing PEI concentration and reaches a maximum rate of 30 percent per minute when the PEI and acrylic acid are present in equimolar proportions [15].

The PEI acts as a template whose surface is effectively saturated with monomer at low concentrations of PEI. In this region the rate increases with increasing PEI concentration. When the PEI concentration increases to the point where insufficient monomer is present to maintain saturation of the template, the rate of polymerization decreases sharply as the continuity of the adsorbed monomer is interrupted; i.e., the matrix is not filled. When the PEI is replaced by tetraethylene pentamine, no template effect is observed at 25°C. The addition of a nonpolymerizable acid, α-hydroxyisobutyric acid, produces a large decrease in the rate of polymerization of acrylic acid due to its competition with the monomer for the template sites.

Thermal initiation by azobisisobutyronitrile at 60°C and photoinitiation by $Mn_2(CO)_{10}$–$CHCl_3$ at 25°C give similar results in the PEI–acrylic acid systems [15].

The rate of the persulfate-initiated polymerization of acrylic acid in aqueous solution at 74°C in the presence of an equimolar amount of poly(ethylene oxide) (PEO) of 6,000 molecular weight increases seven times over the rate in the absence of PEO. However, the increase in rate is attributed to the system becoming heterogeneous, since there is only a slight increase in rate when the reaction is carried out in the homogeneous system that results when the PEO has a molecular weight of 1,500 [14]. Nevertheless, the tacticity of the PAA obtained by persulfate-initiated polymerization of acrylic acid in the presence of PEO at 60°C is different from that obtained in the absence of PEO [12].

The tacticity of the PAA obtained by persulfate-initiated polymerization of acrylic acid in the presence of poly(methyl vinyl ether) at 70°C is also different from that of the normal radical polymer [12]. However, atactic and isotactic polystyrene, poly(methyl methacrylate), and poly(2-vinylpyridine) have no apparent effect on the tacticity of the PAA obtained by benzoyl peroxide-initiated polymerization of acrylic acid in their presence [12].

Polymer–Monomer Interaction Through Stereoregular Fit

The radical-catalyzed polymerization of methyl methacrylate at normal temperatures yields atactic polymer, while low-temperature polymerization—e.g., under ultraviolet light at −60°C in the presence of a photosensitizer—yields syndiotactic polymer. Polymerization initiated by an organolithium compound or an organomagnesium bromide at 25°C yields isotactic poly(methyl methacrylate) (PMMA). These stereoregular structures are shown in Equation (5):

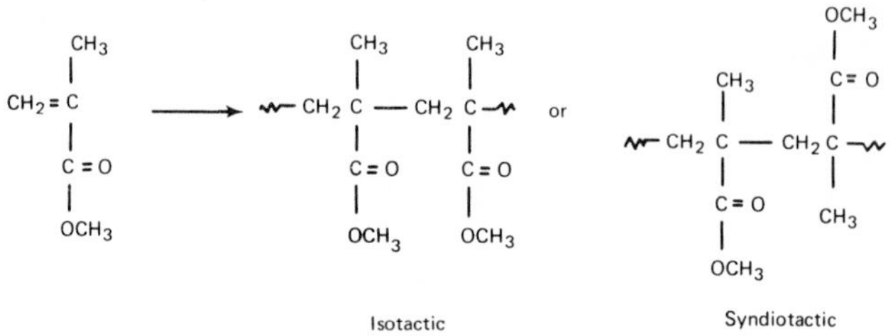

When the polymerization of MMA is conducted at −50°C in the presence of an organomagnesium chloride, the product has high contents of both isotactic and syndiotactic sequences. Although this product was originally presumed to be a "stereoblock" polymer, it was subsequently demonstrated that the same product could be obtained by mixing isotactic and syndiotactic polymers produced as indicated earlier. The product was then identified as a 1:2 (iso/syndio) stereocomplex. The 1:1 (iso/syndio) complex is formed initially and then associates with syndiotactic polymer.

The role of the stereocomplex in the polymerization of MMA is demonstrated when the reaction is carried out with n-butyl magnesium chloride in toluene at −50°C in the presence of either isotactic or syndiotactic PMMA [16]. The presence of the isotactic PMMA promotes the polymerization of MMA to the syndiotactic polymer, while the presence of syndiotactic PMMA results in the formation of isotactic polymer. Thus, the stereospecific polymerization of MMA is closely linked to the formation of the stereocomplex.

The polymerization of MMA in dimethylformamide in the presence of isotactic PMMA (MMA/PMMA = 6) using bis(t-butyl cyclohexyl percarbonate) as initiator at 25°C yields syndiotactic PMMA, which becomes associated with the isotactic substrate to form the stereocomplex [17]. The molecular weight (viscosity average) of the syndiotactic polymer that is formed is parallel to the molecular weight of the isotactic polymer as the latter increases from 1.1×10^5 to 7.2×10^5. The syndiotacticity of the PMMA decreases with conversion and is highest in the presence of the highest-molecular-weight isotactic PMMA. At high conversions additional isotactic PMMA is formed.

Polymerization of MMA in the presence of preformed syndiotactic PMMA results in the formation of isotactic PMMA, while polymerization in the presence of preformed stereocomplex yields additional stereocomplex.

The polymerization is presumed to involve initial growth of polymer radicals in the solution or in the vicinity of the polymer matrix. After reaching a certain DP, the tactic macroradical complexes with the matrix polymer, and propagation then continues in direct contact with the latter. The formation of additional isotactic PMMA at high conversion indicates that the syndiotactic PMMA formed in the presence of isotactic PMMA also operates as a matrix and, in turn, induces the formation of new isotactic PMMA. Although graft polymerization is negligible at 25°C, it may be more significant at higher temperatures [17].

Polymer–Monomer Interaction Through Charge-Transfer Complexation

The interaction of a strong electron donor monomer with an organic electron acceptor, which results in the formation of a charge-transfer complex, is followed by spontaneous or photoactivated polymerization, e.g., N-vinylcarbazole-tetranitromethane. Similarly, the interaction of an organic electron donor with a strong electron acceptor monomer, which leads to the formation of the complex, is also followed by its subsequent polymerization, e.g., pyridine–maleic anhydride. When both the electron donor and the electron acceptor are polymerizable monomers, the formation of the charge-transfer complex and its spontaneous or catalyzed homopolymerization results in the formation of an equimolar, alternating copolymer, e.g., styrene–maleic anhydride or α-olefin-sulfur dioxide.

Poly(maleic anhydride) (PMAnh) undergoes strong charge-transfer interaction with 2- and 4-vinylpyridines. In the presence of PMAnh, the vinylpyridines polymerize spontaneously at 50°C. When DMF is used as solvent, the system remains homogeneous, the polymer yield is quantitative, and the product contains equimolar amounts of PMAnh and poly(vinylpyridine). The molecular weight of the poly(4-vinylpyridine), isolated by solvent extraction, is comparable to that of the PMAnh charged. No polymerization occurs when 4-vinylpyridine is added to succinic anhydride in DMF. It is presumed that alignment of the monomer units on the PMAnh matrix results in charge-transfer interaction, followed by polymerization [18]. Since the residue from the solvent extraction presumably was not examined, there is no indication whether grafting occurs.

A mixture of 4-vinylpyridine and p-chlorostyrene undergoes spontaneous polymerization in the presence of PMAnh in DMF at 50°C. The rate of polymerization is slower than in the absence of the chlorostyrene, and the product is very rich in 4-vinylpyridine. Neighboring vinylpyridine units on the PMAnh matrix apparently undergo polymerization while the comonomer enters the copolymer to a very limited extent [19].

4-Vinylpyridine polymerizes rapidly in the presence of an equimolar styrene-maleic anhydride copolymer in DMF at 50°C. The product contains equimolar amounts of the matrix copolymer and poly(vinylpyridine). The rate of polymerization is slower than that in the presence of PMAnh, indicating the effect of the separation of anhydride units by styrene units [19].

The complexation of a relatively weak electron acceptor monomer such as acrylonitrile or methyl methacrylate with a metal halide or organoaluminum halide increases the electron-accepting capability of the monomer and permits the formation of a comonomer charge-transfer complex by interaction with an electron-donating monomer. The comonomer complex undergoes spontaneous or catalyzed homopolymerization to yield an equimolar, alternating copolymer, irrespective of initial monomer charge [20].

$$A + MX \longrightarrow A...MX \tag{6}$$

$$D + A...MX \rightleftharpoons \left[D \rightarrow A...MX \right] \rightleftharpoons \left[D^{+} \overset{\cdot}{\cdot} A \; ...MX \right] \tag{7}$$

$$x \left[\begin{matrix} D^{+} \overset{\cdot}{\cdot} A \\ \vdots \\ MX \end{matrix} \right] \longrightarrow \left[\begin{matrix} D - A \\ \vdots \\ MX \end{matrix} \right]_{x} D^{\pm} \overset{\cdot}{\cdot} A \\ MX \tag{8}$$

During the initial, rapid stage of polymerization, the molecular weight of the copolymer increases with conversion. In the second, slower stage of polymerization, the molecular weight remains at the maximum level attained in the initial stage. It has been proposed [21] that the comonomer complexes are arranged in a matrix or ordered array, as shown in Figure 1.

The molecular weight of the copolymer formed in the first stage of the polymerization is determined by the size of the matrix, i.e., the complex concentration. The diffusion of the unpolymerized monomers to the complexing agent affixed to the copolymer chain results in the generation of new complexes on or in the immediate vicinity of the copolymer. The latter acts as a template, and the slow rate of polymerization is a function of the time necessary for the formation of a void-free array of complexes. The new polymer replicates the molecular weight of the original matrix. This process is repeated until full conversion is attained.

When a mixture of the comonomers is added to the system at full conversion or at any conversion after the molecular weight has become constant, the added

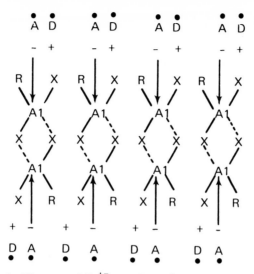

Figure 1. Alignment of $[D \overset{+-}{\cdot\cdot} A \ldots RAlX_2]$ complexes in matrix.

monomers yield a copolymer with the same molecular weight as the initial copolymer at all conversions during their polymerization [21].

As indicated in Equation (8), the propagating chain-end in the homopolymerization of a comonomer charge-transfer complex is a complex. It has been proposed [20] that termination involves the formation of a carbene that is inserted into a tertiary —CH to generate an olefin or a six-membered ring.

Alternating copolymer graft copolymers are produced when comonomers that form charge-transfer complexes polymerize in the presence of a polymer containing tertiary —CH groups [22]. Grafting occurs by termination of a propagating alternating copolymer chain by insertion of the polymeric carbene into the labile —CH of the substrate polymer, as shown in Equation (9):

(9)

TABLE II

Graft Copolymerization of Poly(styrene-alt-acrylonitrile) onto Nitrile Rubber
Precomplexed with Ethylaluminum Sesquichloride (EASC)

Precomplexation		Polymerization	
P(B-AN) (70:30), g	4.0	AN, g (mmoles)	4.1 (77)
EASC, mmoles	22.0	S, g (mmoles)	8.0 (77)
Toluene, ml	40.0	Benzoyl peroxide, g	0.36
Temp./time, °C/hr	40/1	Temp./time, °C/hr	40/5

Monomer Conversion 87%

Fractionation

Solvent	%	B/AN/S mole ratio	S/AN mole ratio
Benzene	19	64/27/9	9/0
Acetone	12	0/50/50	50/50
Residue	69	40/39/21	49/51

High levels of grafting are attained on polymers containing polar functionality, e.g., ester or nitrile groups, when the latter are complexed with a complexing agent before the addition of the comonomers [22,23]. The complexing agent is anchored onto the substrate polymer, and the comonomer complexes are therefore generated in the vicinity of the polymer. Polymerization of the com-

TABLE III

Graft Copolymerization of Poly(styrene-alt-acrylonitrile) onto Poly(butyl acrylate)
(PBA) Latex Precomplexed with $ZnCl_2$

Precomplexation		Polymerization	
PBA latex (42.5%), g	18.8	AN, g (mmoles)	6.6 (125)
PBA, g (mmoles)	8.0 (62.5)	$K_2S_2O_8$, g	0.4
Styrene, g (mmoles)	13.0 (125)		
Temp./time, °C/hr	30/24	Temp./time, °C/hr	50/20
$ZnCl_2$, g (mmoles)	4.26 (31.25)		
Temp./time, °C/hr	42/5		

Monomer Conversion 88.4%

Fractionation

Solvent	%	BA/S/AN mole ratio	S/AN mole ratio
Hexane	8.7	100/0/0	–
Acetone	7.3	1/46.5/52.5	47/53
Residue	84.0	24/40/35	53/47

plexes results in the formation of alternating copolymer chains that terminate by insertion into the substrate polymers.

The graft copolymerization of poly(styrene-alt-acrylonitrile) onto nitrile rubber precomplexed with ethylaluminum sesquichloride is shown in Table II, while grafting onto poly(butyl acrylate) latex precomplexed with zinc chloride is shown in Table III.

When the copolymerization of styrene and acrylonitrile in the presence of zinc chloride is carried out in the presence of cellulose, under conditions that normally yield an alternating copolymer, i.e., in an aqueous system in the presence of a water-soluble radical precursor at a temperature below 60°C, the alternating copolymer is accompanied by cellulose graft copolymer in which the grafted chains have an equimolar, alternating structure [24]. The monomer conversion is higher than when the reaction is carried out in the absence of cellulose, and both grafted copolymer (60–80 percent of the total copolymer) and ungrafted copolymer are of extremely high molecular weight (Table IV).

The comonomer complexes $[S \overset{\text{\tiny +}}{\cdots} AN \ldots ZnCl_2]$ may be anchored on the cellulose through the interaction of $ZnCl_2$ and the cellulosic hydroxyl groups, since aqueous solutions of the metal halide are known to break the hydrogen bonds in cellulose and reduce the crystallinity. Although radical sites generated on the cellulose as a result of attack by initiator radicals may initiate polymerization of the comonomer charge-transfer complexes, they are not the sites for the attachment of grafted alternating copolymer chains since catalyst residues are not usually found in alternating copolymers. The grafting sites are probably

TABLE IV

Graft Copolymerization of Styrene and Acrylonitrile
onto Cellulose in the Presence of $ZnCl_2$

Styrene, moles	1.0	1.0
Acrylonitrile, moles	1.0	1.0
$ZnCl_2$, moles	0.5	0.5
$K_2S_2O_8$, moles	0.05	0.05
Water, g	157.0	157.0
Wood pulp, g	10.0	10.0
Temp./time, °C/hr	40/5	50/1
Monomer conversion, %	32	16
Add-on, %	353	146
Grafted copolymer		
% of total	66.7	58.6
AN, mole-%	49.0	50.2
$[\eta]$, dl/g (DMF, 30°C)	5.3	5.2
Ungrafted copolymer		
% of total	33.3	41.4
AN, mole-%	50.2	52.0
$[\eta]$, dl/g (DMF, 30°C)	4.3	3.2

aldehyde groups generated on the cellulose by cleavage of the terminal hemiacetal groups, and grafting occurs by insertion of the propagating chains into the aldehydic —CH group [25]. Thus, the cellulose graft copolymers are actually block copolymers.

When methyl methacrylate is heated with cellulose fibers (cotton, rayon, hemp) at 90°C in an aqueous suspension in the absence of a radical initiator, graft polymerization readily occurs to give an approximately 100 percent add-on on the fibers. In addition, high-molecular-weight poly(methyl methacrylate) is produced simultaneously in high yield, and the total monomer conversion may reach 100 percent. Under similar conditions, grafting on silk yields as much as 1,000 percent add-on, and the amount of homopolymer at full monomer conversion is correspondingly reduced. A lesser amount of grafting occurs on wool and polyacrylonitrile fibers, although homopolymerization occurs to the extent of 11–40 percent. Homopolymerization is accompanied by a small amount of grafting on nylon.

The grafting and homopolymerization of methyl methacrylate occur to the indicated extent in the presence of the fibers, under conditions that yield negligible amounts of polymer in the absence of the fibers [26].

Styrene and acrylonitrile are not grafted to cellulose fibers under conditions that are effective for methyl methacrylate, while methyl acrylate, styrene, and glycidyl methacrylate are grafted to silk, although to a very small extent.

The graft polymerization of methyl methacrylate on cellulose and silk in the absence of initiator at 85°C can be carried out in water or a water–acetone mixture but does not occur in the absence of water, i.e., in pure methanol, acetone, benzene, dioxane, dimethylformamide, or tetrahydrofuran. The polymerization takes place in the dark as well as in the light [27].

The presence of hydroxyl or amide groups and their location and steric arrangement all play an important role in the initiation mechanism. The polymers

TABLE V

Effective Substrate Polymers for Uncatalyzed
Polymerization of Methyl Methacrylate in
Presence of Water

Cotton	Starch
Rayon	Dialdehyde starch
Hemp	Wool
Paper	Silk
Wood pulp	Nylon
Polyvinyl alcohol, partially formalized	
Polyvinyl acetate, partially hydrolyzed	
Poly(methyl methacrylate–co–hydroxyethyl methacrylate)	
Poly(methyl methacrylate–co–acrylic acid)	
Poly(styrene–co–allyl alcohol)	

shown in Table V effectively initiate the homopolymerization and/or the graft copolymerization of methyl methacrylate in the absence of an added initiator. Low-molecular-weight polyols and low- or high-molecular-weight commercial poly(vinyl alcohol) and poly(vinyl acetate) are ineffective [28].

The mechanism proposed for the polymerization of methyl methacrylate in the presence of cellulose or other effective polymers and water involves the formation of a complex between the polymeric −OH (or −CONH− in the case of silk) groups, water and monomer, as shown in Equation (10):

(10)

The presence of carbon tetrachloride has a dramatic effect on the cellulose–water–vinyl monomer system [29]. In the presence of a small amount of CCl_4, the conversion of monomer to polymer after 5 hours at 85°C increases rapidly up to a quantitative yield and then decreases as the CCl_4 content is increased further. The conversion is zero in the absence of water. In contrast, in the absence of CCl_4, the conversion after 5 hours at 85°C is 10 percent. Methyl acrylate also readily polymerizes at 85°C in the presence of cellulose, water, and CCl_4, while the failure of acrylonitrile and styrene to polymerize in the presence of cellulose and water is not influenced by the presence of CCl_4. N-vinyl-2-methylimidazole polymerizes in the cellulose–water–CCl_4 system at 30°C, while the acrylic monomers are unpolymerized after 5 hours at 30°C.

Although the conversion increases in the presence of CCl_4, the grafting efficiency decreases exponentially with increasing amounts of CCl_4 in the cellulose–water–methyl methacrylate system at 85°C. The molecular weight of the homopolymer decreases with increasing CCl_4 content, apparently due to conventional chain transfer.

The presence of CCl_4 also influences the methyl methacrylate polymerization in the silk–water system at 85°C. Conversion and efficiency of grafting decrease with increasing amounts of CCl_4 in the system. However, in comparison with the cellulose system, the grafting efficiency is quite high, probably due to chain transfer of growing chain-ends to the silk.

Since the grafting efficiency in the cellulose–water–CCl_4 system is lower than that in the cellulose–water system, the proposed reaction mechanism for initiation of methyl methacrylate polymerization has been modified, as shown in Equation (11):

$$(11)$$

According to this mechanism, the formation of graft copolymer is due predominantly to chain transfer of homopolymer chains. Cellulose is therefore a part of the complex initiator and acts as a polymerization matrix.

Not all types or samples of cellulose are equally effective in initiating the polymerization of methyl methacrylate at 85°C. However, the efficiency is greatly promoted on the addition of a small amount of CCl_4, although no polymerization occurs in the absence of either water or cellulose.

Cuprammonium rayon fiber is even more effective than wood pulp in the cellulose–water–CCl_4–methyl methacrylate system at 85°C. The promoting effect of CCl_4 in the cellulose–water–methyl methacrylate system is not demonstrated when styrene or acrylonitrile are substituted for the methyl methacrylate, even in the presence of the very effective cuprammonium rayon fiber [30].

The addition of cupric ion to a mixture of methyl methacrylate and cotton in the presence of water greatly increases the polymer yield in the absence of CCl_4 [31]. It has been proposed that cupric ion plays an important role in the presence of cellulose–water, and the mechanism in Equation (10) has been revised, as shown in Equations (12) and (13):

$$(12)$$

$$(13)$$

The use of ceric salts is generally considered to be one of the most effective methods of promoting the formation of graft copolymers on cellulose. It is presumed that a ceric ion–cellulose complex is initially formed; then, as a result of a one-electron transfer, the ceric ion is reduced to cerous ion, and a free radical is generated directly on the cellulose backbone. The grafting site is considered to be the C_2-C_3 glycol group of the anhydroglucose unit, which undergoes ceric ion–induced cleavage to generate a carbonyl group and a radical that initiates graft copolymerization, as shown in Equation (14):

$$(14)$$

It has also been proposed that the site of attack is the hemiacetal group of the end unit of the cellulose molecule. Analogous to the situation in which the C_2-C_3 glycol group is cleaved, the hemiacetal is converted into a radical and a carbonyl group, as shown in Equations (15) and (16). However, in contrast to the cleavage of the glycol unit that leads to a graft copolymer, the cleavage of the hemiacetal leads to a block copolymer.

$$(15)$$

$$(16)$$

It has recently been proposed [25,32] that cellulose–monomer interaction or complexation plays a major role in the grafting of acceptor monomers such as acrylic esters, amide, or nitrile on cellulose in the presence or absence of catalysts, and that the behavior of the cellulose-monomer complex is analogous to that of donor-acceptor complexes generated from acceptor monomers or donor monomer-acceptor monomer pairs in the presence of metal halides or other Lewis acids [21].

The major features of the proposed mechanism for the graft copolymerization of polar acceptor monomers on cellulose are as follows:

1. Cellulose acts as a complexing agent for the activation of acceptor monomers and as a matrix for the alignment of monomer-cellulose-water-donor-acceptor complexes, as shown in Equations (17) and (18):

$$
\begin{array}{c}
\text{X--C=O} \quad\text{X--C=O}\quad \text{X--C=O}\quad \text{X--C=O} \\
| \qquad\qquad | \qquad\qquad | \qquad\qquad | \\
\text{R--C}^+ \;\;{}^-\text{CH}_2 \quad \text{R--C}^+\;{}^-\text{CH}_2 \quad \text{R--C}^+\;{}^-\text{CH}_2 \quad \text{R--C}^+\;{}^-\text{CH}_2 \\
|\quad\;| \qquad |\quad\;| \qquad |\quad\;| \qquad |\quad\;| \\
\text{H}_2\text{C}^\cdot \;\;{}^\cdot\text{C--R} \;\; \text{H}_2\text{C}^\cdot\;{}^\cdot\text{C--R}\;\;\text{H}_2\text{C}^\cdot\;{}^\cdot\text{C--R}\;\;\text{H}_2\text{C}^\cdot\;{}^\cdot\text{C--R} \\
\text{X--C} \qquad \text{X--C} \qquad \text{X--C} \qquad \text{X--C}
\end{array}
\tag{17}
$$

$$
\begin{array}{c}
\text{CN}\qquad\qquad\text{CN}\qquad\qquad\text{CN}\qquad\qquad\text{CN} \\
|\qquad\qquad|\qquad\qquad|\qquad\qquad| \\
\text{R--C}^+\;\;{}^-\text{CH}_2\;\;\text{R--C}^+\;{}^-\text{CH}_2\;\;\text{R--C}^+\;{}^-\text{CH}_2\;\;\text{R--C}^+\;{}^-\text{CH}_2 \\
|\quad\;|\qquad|\quad\;|\qquad|\quad\;|\qquad|\quad\;| \\
\text{H}_2\text{C}^\cdot\;\;{}^\cdot\text{C--R}\;\;\text{H}_2\text{C}^\cdot\;{}^\cdot\text{C--R}\;\;\text{H}_2\text{C}^\cdot\;{}^\cdot\text{C--R}\;\;\text{H}_2\text{C}^\cdot\;{}^\cdot\text{C--R}
\end{array}
\tag{18}
$$

When ceric ion is present, cellulose-monomer-water and ceric ion-monomer-water complexes are present. The complex containing ceric ion is anchored on the cellulose matrix by interaction of the ceric ion with the carbonyl groups present thereon initially or generated by the cellulose-ceric ion redox reaction, as shown in Equation (19):

$$\begin{array}{cccc}
\text{X-C=O} & \text{X-C=O} & \text{X-C=O} & \text{X-C=O} \\
| & | & | & | \\
\text{R-C}^+ \ ^-\text{CH}_2 & \text{R-C}^+ \ ^-\text{CH}_2 & \text{R-C}^+ \ ^-\text{CH}_2 & \text{R-C}^+ \ ^-\text{CH}_2 \\
| \quad | & | \quad | & | \quad | & | \quad | \\
\text{H}_2\text{C} \cdot \ \cdot \text{C-R} & \text{H}_2\text{C} \cdot \ \cdot \text{C-R} & \text{H}_2\text{C} \cdot \ \cdot \text{C-R} & \text{H}_2\text{C} \cdot \ \cdot \text{C-R} \\
| & | & | & | \\
\text{X-C} & \text{X-C} & \text{X-C} & \text{X-C}
\end{array}$$

(19)

2. Homopolymerization of the monomer complexes is initiated spontaneously, i.e., thermally, if the complex concentration is high, or under the influence of oxygen, light, radicals from the catalyst or radicals generated on the cellulose as a result of swelling or cellulose–catalyst interaction or complexation.

3. The initiation step involves hydrogen abstraction from the monomer in the donor–acceptor complex and, since the abstracting radical does not become incorporated into the polymer chain, initiation by radicals on the cellulose does not result in grafting, to any appreciable extent.

4. Polymer chains that ultimately become grafted ("grafted copolymer") or remain ungrafted ("homopolymer") are both, for the most part, derived from monomer–cellulose–water complexes and are initiated and propagated by the same nonradical mechanism. However, cellulose radicals and catalyst radicals may initiate some graft copolymerization and homopolymerization, respectively, of uncomplexed monomer. The polymerization of uncomplexed monomer proceeds by a radical mechanism, and some homopolymer may therefore result from chain transfer to monomer. The polymerization of complexed monomer involves nonradical propagating chain-ends that do not participate in chain-transfer reactions. However, at elevated temperatures, the chain-end becomes a radical that propagates and terminates in a conventional manner, e.g., disproportionation, and undergoes chain transfer. Similarly, at high dilution, e.g., high water content, or at high conversions, when complex concentration is reduced, the chain-end becomes a radical species and behaves in a conventional manner. Since the polymerization of the complexed monomer is initiated without incorporation of initiating radicals, chain-ends that become radical species under the indicated conditions result in the formation of ungrafted homopolymer.

5. Propagating chains from the polymerization of complexed monomer terminate on the cellulose by insertion into aldehyde —CH bonds and become grafted copolymer, while chains that do not find suitable insertion sites or that undergo prior termination by insertion into monomer or polymer become ungrafted

homopolymer. Due to the crystallinity of cellulose, the complexes are essentially aligned on the surface of the cellulose matrix, and the propagating chains terminate in the more accessible amorphous areas, e.g., on aldehyde groups generated by cleavage of the terminal hemiacetal groups of the cellulose chains whose ends are in the amorphous areas. The high molecular weights of the grafted copolymer and the homopolymer are the result of the alignment of the complexes on the cellulose, the absence of chain transfer and other characteristics of radical chain-ends, and the limited accessibility of suitable terminating insertion sites on the cellulose.

The structures shown in Equations (17)–(19) are analogous to the structures proposed in the homopolymerization of acceptor monomers in the presence of metal halides, since complexation of an acceptor monomer results in delocalization of electrons in the double bond and increases its electron-accepting ability. As a result, uncomplexed monomer becomes a donor relative to the complexed monomer, and a donor–acceptor complex is generated. The greater the electron-donating ability of the uncomplexed acceptor monomer, the greater the tendency to form donor–acceptor complexes. The higher the complex concentration, the higher the rate of polymerization and the molecular weight of the polymer, and the greater the extent of grafting.

The effect of cellulose-monomer interaction is clearly shown in the polymerization of methacrylonitrile using ceric ion or ammonium persulfate as initiator [32]. Although there is low or negligible monomer conversion in the absence of cellulose, as shown in Table VI, there is a dramatic increase in conversion, under the same conditions, in the presence of cellulose. In fact, at least a fortyfold increase in conversion is noted in the presence of ceric ion even with a reduction in reaction time from 3–6 hours to 1.25 hours. The increase in conversion in the presence of the persulfate catalyst, while not as dramatic, is still highly significant.

TABLE VI

Graft Copolymerization of Methacrylonitrile onto Cellulose*

Cellulose, g	0	5	0	5	5
Catalyst, mmoles	CAN, 1	CAN, 1	APS, 5	APS, 5	APS, 5
Temp., °C/time, hr	40/6	40/1.25	50/3	50/3	50/22
Conversion, %	<1	35.4	<1	6.1	82.2
Polymer					
Ungrafted, %	100	9	100	64	4
$[\eta]^\dagger$	0.15	2.7	0.28	3.0	4.3
Grafted, %		91		36	96
$[\eta]^\dagger$		3.2		5.3	6.0
Add-on, %		215		16	527

*MAN, 33.5 g (0.5 mole); H_2O, 38.5 g.
†dl/g in DMF at 25°C.

The molecular weight of the polymer resulting from the polymerization of donor-acceptor complexes increases with conversion up to a limiting molecular weight that is dependent upon the initial complex concentration and the availability of terminating insertion sites. Since the major part of the homopolymer as well as the grafted copolymer result from complexation, the molecular weight of the homopolymer would be expected to increase with conversion, as shown in Table VI.

The stereoregularity of poly(methyl methacrylate) grafted onto rayon or poly(vinyl alcohol) fibers, using potassium persulfate as initiator, is the same as that of the atactic polymer obtained in the absence of the fibers. However, the stereoregularity of poly(methyl acrylate) grafted onto poly(vinyl alcohol) fibers is distinctly different. Rayon has no stereoregulating effect during the grafting of methyl acrylate [33].

Since vinyl acetate has an ester carbonyl group similar to that of the acrylic esters, it should also be capable of forming complexes of the type shown in Equation (17). The grafting of vinyl acetate onto rayon fibers, with potassium

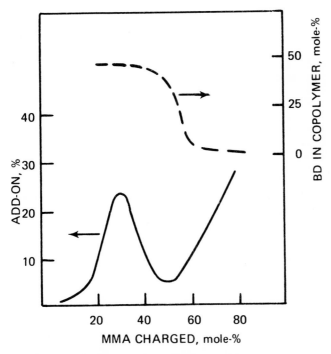

Figure 2. Butadiene content and percent add-on of grafted copolymer in uncatalyzed copolymerization of butadiene and methyl methacrylate in the presence of kraft wood pulp as a function of monomer charge after 5 hours at 90°C.

persulfate as initiator, has been found to yield grafted polymer containing a considerable amount of head-to-head structure [33].

The ability of cellulose to act as a complexing agent for the activation of acceptor monomer and as a matrix for the alignment of comonomer complexes [$D^{+.-}$ A–cellulose–water] has been demonstrated by the formation of equimolar, alternating copolymers from binary monomer mixtures in the absence as well as in the presence of catalysts.

When a mixture of butadiene and methyl methacrylate is heated at 90°C with an aqueous suspension of kraft wood pulp containing a small amount of nonionic surfactant, in the absence of a catalyst, essentially equimolar alternating butadiene–methyl methacrylate copolymer is grafted on the cellulose when the monomer charge contains 20–50 mole-percent methyl methacrylate, as shown in Figure 2 [34].

The copolymerization of styrene and methyl methacrylate (90/10 molar ratio) in the presence of wood pulp at 90°C, using $NaClO_2$ as catalyst, results in the formation of equimolar grafted and ungrafted copolymers at a low conversion. However, while the grafted copolymer composition is still equimolar at a higher conversion, the ungrafted copolymer composition approaches that expected from the comonomer charge in accordance with a radical mechanism (Table VII) [35].

TABLE VII

Graft Copolymerization of Styrene–Methyl Methacrylate
onto Wood Pulp in Presence of $NaClO_2$

Charge S/MMA mole ratio	90/10	90/10
Temperature, °C	90	90
Time, hr	10	48
Conversion, %	16	61
Add-on, %	9	45
Copolymer S/MMA mole ratio		
Grafted	48/52	45/55
Ungrafted	52/48	89/11

The proposed alignment of comonomer complexes on the cellulose matrix is shown in Equation (20) on the following page.

The mutual γ-irradiation of viscose rayon immersed in a methanol solution containing butadiene and acrylonitrile at 30°C results in the formation of ungrafted copolymers having the compositions expected for a normal radical copolymerization over a wide range of monomer charge ratios. However, the grafted copolymers are essentially equimolar when the acrylonitrile content in the monomer charge ranges from 30–90 mole-percent [36].

Under similar conditions, the graft polymerization of a mixture of styrene

(20)

and acrylonitrile on viscose rayon yields equimolar grafted copolymers and un-grafted copolymers with a radical composition when the monomer charge contains 25-75 mole-percent acrylonitrile.

Similar results are obtained in the radiation-induced graft copolymerization of butadiene-acrylonitrile and styrene-acrylonitrile onto poly(vinyl alcohol) fibers [36].

Alternating styrene-acrylonitrile and styrene-methacrylonitrile copolymers are produced both as ungrafted and cellulose graft copolymers when the copolymerizations are carried out at 40-50°C in the presence of cellulose, using a persulfate, a persulfate-bisulfite redox system, or ceric ammonium nitrate as initiator [37]. The conditions and results are shown in Tables VIII and IX.

TABLE VIII

Grafting of Styrene–Methacrylonitrile onto Cellulose*

Cellulose, g	0	5	0	5
S + MAN, moles	0.50	1.0	0.50	1.0
Water, g	42.75	90.5	42.75	90.5
Catalyst, mmoles	APS, 10	APS, 5	CAN, 1	CAN, 2
Temp., °C/time, hr	40/3	50/3	40/3	40/3
Conversion, %	3.7	1.9	0.7	2.9
Polymer				
Ungrafted, %	100	65	100	51
AN, mole-%	44	51		53
$[\eta]$ dl/g†	0.6	1.1	ins	1.0
Grafted, %		35		49
AN, mole-%		52		53
$[\eta]$ dl/g†		2.7		1.6
Add-on, %		11		25

*S/MAN mole ratio, 1; H_2O, 50 wt-%.
†DMF, 30°C.

TABLE IX

Grafting of Styrene–Acrylonitrile onto Cellulose*

Cellulose, g	0	5	0	5
S + AN, moles	0.40	1.0	0.50	1.0
Water, g	31.4	83.5	39.25	83.5
Catalyst, mmoles	APS, 10		CAN, 1	CAN, 2
Catalyst, g		KPS/SBS 0.05/0.02		
Temp., °C/time, hr	40/3	30/3	40/3	40/3
Conversion, %	3.2	3.6	5.6	15.0
Polymer				
Ungrafted, %	100	57	100	33
AN, mole-%	27	44	26	47
$[\eta]$ dl/g†	2.5	3.7	1.4	3.5
Grafted, %		43		67
AN, mole-%		48		48
$[\eta]$ dl/g†		6.5		4.2
Add-on, %		25		158

*S/AN mole ratio, 1; H_2O, 50 wt-%.
†DMF, 30°C.

Charge-transfer interactions have been proposed in the radiation-induced grafting of monomers on cellulose [38–40]. "Inaccessible" radical sites generated on the cellulose by irradiation form charge-transfer complexes with adsorbed monomer, e.g., styrene, as a result of the delocalization of the styrene π-electrons into the free valences of the irradiated cellulose. Thus, where $\dot{P}$ is a radical site either on the surface or within the cellulose, as shown in Equation (21), the monomer is held or adsorbed horizontally through the aromatic ring π-electrons:

$$(21)$$

Rotation of the monomer molecule results in a change from the π-bonding to the vertical σ-bonding, as shown in Equation (22):

$$(22)$$

Bonding through the side chain occurs in an analogous manner:

In addition, consistent with the observed difficulty in removing ungrafted homo-polymer from the bulk mixture with cellulose, long polystyrene chains may be held by a series of π-bonds:

Before leaving the subject of cellulose interaction with monomers, it is of interest to note that lignin polymers are produced at room temperature from the reaction of eugenol, peroxidase, and hydrogen peroxide in aqueous medium in plant tissue slices and cell walls, but not in the absence of the wall material. Lignin is also formed when the reaction is carried out in the presence of filter paper, starch, cotton, methylcellulose, milkweed fiber, deacetylated chitin, and arabic acid. Thus, polysaccharides can serve as matrices for the peroxidase-catalyzed polymerization of eugenol to lignin polymers [41,42].

As indicated earlier, methyl methacrylate undergoes homopolymerization and forms grafted copolymers when heated in an aqueous medium in the presence of silk fibers in the absence of a radical initiator [26,27,43,44]. The monomer conversion and grafting efficiency are much higher than when the reaction is carried out in the presence of cellulose.

The poly(methyl methacrylate) grafted on silk, using $LiBr-K_2S_2O_8$ as initiator, has a higher isotacticity than the polymer produced in the absence of the silk fibers, suggesting a stereoregulating effect on the structure of the polymer formed within the fibers [45].

The polymerization of methyl methacrylate occurs readily in the presence of wool fibers in the absence of initiators [46]. Grafting is also accomplished when the reaction is initiated by ceric salts [47] or the $LiBr-K_2S_2O_8$ system [45]. In the latter case the grafted polymer appears to have a higher tacticity than normally atactic polymer.

The polymerization of methyl methacrylate proceeds very readily in the presence of collagen in the absence of a radical initiator. The grafting efficiency is influenced by the water content of the reaction mixture, and the conversion is much higher than that obtained in the presence of cotton or starch [48].

The polymerization of methyl methacrylate at 85°C in the presence of nylon-6 and water and in the absence of an initiator results in the formation of homo-polymer and graft copolymer. The nylon–water system does not initiate the polymerization of styrene, acrylonitrile, methyl acrylate, or vinyl acetate. The polymerization occurs predominantly inside the nylon fiber, and the molecular weight of the homopolymer formed inside the fiber is considerably higher than that formed outside the fiber [49,50]. The higher the content of amine end-groups in the nylon molecule, the lower the monomer conversion. The presence of a small number of cupric ions is necessary for the initiation of polymeriza-tion. Treatment of the nylon with formic acid removes the cupric ion and de-creases the initiating ability. The addition of a trace of cupric salt to the nylon regenerates its activity. The inhibition and kinetic behavior suggests a radical mechanism in which cupric ions and monomer are adsorbed or complexed on the nylon, as shown in Equation (25) [51] :

$$(25)$$

Methyl methacrylate undergoes rapid and quantitative polymerization at 85°C when CCl_4 is present in the nylon–water system. The conversion is independent of the amine end-group content in the nylon-6. The radical nature of the re-action is suggested by air retardation, hydroquinone inhibition, and copolymer composition from styrene–methyl methacrylate at 85°C. The proposed [51] mechanism of polymerization in the presence of CCl_4 is shown in Equation (26):

$$(26)$$

The polymer obtained in the presence of nylon–water–cupric ion is the same as that obtained by normal radical polymerization [52]. Similarly, poly(methyl methacrylate) grafted onto nylon-6 fibers by persulfate-catalyzed polymerization in methanol–water at 60°C has the same stereostructure as normal atactic poly-mer. However, poly(methyl acrylate) and polystyrene grafted on nylon-6 under the latter conditions have distinctly different stereostructures from the normal atactic polymers [53]. Poly(vinyl acetate) grafted on nylon-6 with persulfate in water on water-methanol at 90°C also has a different stereoregularity than normal poly(vinyl acetate), although the chemical structures appear to be the same [54].

Polymer–Monomer Interaction Through Unspecified Mode

As indicated earlier, N-vinyl-2-methylimidazole readily polymerizes at 30°C in the cellulose-water-CCl$_4$ system [29,55]. The product consists mainly of homopolymer accompanied by a small amount of cellulose graft copolymer. The reaction does not occur in the absence of cellulose, water, or CCl$_4$, in contrast to the polymerizability of methyl methacrylate in the absence of CCl$_4$. The polymerization is inhibited by hydroquinone, and chlorine is incorporated in the polymer. Silk and poly(vinyl alcohol) also initiate the polymerization at 30°C in the presence of CCl$_4$, but lower alcohols are ineffective. Chloroform and methylene chloride cannot be substituted for CCl$_4$ in the cellulose-water system. Carbon tetrabromide gives a lower yield of polymer than CCl$_4$ [55,56].

In the cellulose-water-CCl$_4$ system at 30°C 2-vinylpyridine, 4-vinylpyridine, 5-ethyl-2-vinylpyridine, and 2-methyl-5-vinylpyridine polymerize only to a very small extent, while methyl methacrylate, methyl acrylate, acrylonitrile, and styrene are not polymerized at all under these conditions [56].

It is apparent that the imidazole moiety plays a role in the initiation process. This is confirmed by the finding that the polymerizations of acrylonitrile, methyl methacrylate, methyl acrylate, acrylic acid, methacrylic acid, and styrene are initiated at 80°C in benzene in the presence of imidazole and CCl$_4$, although cellulose and water are absent. Vinyl chloride and methyacrylonitrile are not polymerized under these conditions. In spite of the presence of CCl$_4$, the molecular weights of the polymers are very high, e.g., up to 2,000,000. While carbon tetrabromide can replace CCl$_4$ at 80°C, chloroform and methylene chloride are ineffective [56, 57].

The copolymerization of acrylonitrile and styrene at 85°C in the imidazole system yields a copolymer whose composition is that of the normal radical copolymer.

In addition to imidazole, the 2-methyl-, 2-ethyl-, 2,4-dimethyl-, and 2-ethyl-4-methylimidazole derivatives are effective in polymerizing acrylonitrile at 80°C in benzene in the presence of CCl$_4$ to high-molecular-weight polymer. Pyrrole, pyridine, 4-ethylpyrimidine, aniline, carbazole, urea, thiourea, histidine, glycine, and valine have little or no initiating ability in the presence of CCl$_4$ [57].

The proposed mechanism of initiation in the imidazole-CCl$_4$ system involves a three-component association, the elimination of hydrogen chloride, and the incorporation of chlorine in the polymer, as shown in Equation (27):

$$\tag{27}$$

Although pyridine and 4-ethylpyrimidine, in contrast to imidazole, do not initiate the polymerization of acrylonitrile at 80°C in benzene in the presence of CCl_4, poly(4-vinylpyrimidine) as well as poly(2-ethyl-N-vinylimidazole) are effective initiators in the presence of CCl_4, while poly(N-vinylcarbazole) and poly(N-vinylpyrrolidone) are ineffective. Since the polymeric amines do not contain —N—H groups, it has been suggested that the mechanism of initiation differs from that in the imidazole–CCl_4 system and may resemble that in the cellulose–water–CCl_4 system [57].

The addition of various polymers in the imidazole–CCl_4 system greatly accelerates the polymerization of acrylonitrile in benzene at 80°C to high-molecular-weight polymer. The effective promoters include cellulose, cellulose–water, poly(4-vinylpyridine), poly(5-methyl-2-vinylpyridine), poly(2-ethyl-N-vinylimidazole), poly(4-vinylpyrimidine), poly(N-vinylcarbazole), poly(N-vinylpyrrolidone), poly(methyl methacrylate) and, to a lesser extent, silk, poly(ethylene oxide), and poly(allyl alcohol). While poly(N-vinylpyrrolidone) and poly-(N-vinylcarbazole) do not initiate acrylonitrile polymerization in the presence of CCl_4 when substituted for imidazole, they act as promoters when added to an imidazole or substituted imidazole–CCl_4 system [57].

The polymerizations initiated in the presence of suitable polymers and imidazole–CCl_4 suggest either that the monomers are anchored on the polymeric matrix and initiation results from anchored monomer–imidazole–CCl_4 interactions or that the imidazole is anchored on the polymeric matrix and increases the localized monomer concentration through monomer–imidazole–CCl_4 interaction.

Although the nature of the interaction is unknown, the stereoregularities of poly(methyl methacrylate) and poly(vinyl acetate) prepared by radical polymerization in the presence of polystyrene and poly(methyl methacrylate) are different from those of the polymers prepared in the absence of the preformed substrate polymers [12].

Summary

The effect of polymer–monomer interactions on the rates of monomer polymerization and graft copolymerization is not too surprising. However, changes in the mechanisms of polymerization and/or graft copolymerization as well as the composition, structure, and stereoregularity of the newly formed polymer provide new opportunities to control the nature and properties of polymers. It is apparent that in polymerizations carried to high conversions, the products possess a new heterogeneity factor, namely, that resulting from a change in environment as the later stages of polymerization take place in the presence of the products of the earlier stages. A new area for investigation has been brought into the light and will inevitably yield new and valuable insights and materials.

References

1. Kabanov, V.A., Aliev, K.V., Kargina, O.V., Patrikeeva, T.I., and Kargin, V.A., "Specific Polymerization of Vinylpyridinium Salts. Polymerization of Macromolecular Matrices," *J. Polym. Sci., Part C*, no. 16, pt. 2 (1967), 1079.

2. Kabanov, V.A., Kargina, O.V., and Petrovskaya, V.A., "Mechanism of the Matrix Synthesis of Polymer–Polymer Salt Complexes," *Polym. Sci. USSR*, 13, no. 2 (1971), 394.

3. Salamone, J.C., Snider, B., and Fitch, W.L., "Polymerization of 4-Vinylpyridinium Salts. III. A Clarification of the Mechanism of Spontaneous Polymerization," *J. Polym. Sci., Part A-1*, 9, no. 6 (1971), 1493.

4. Mielke, I. and Ringsdorf, H., "Investigation of Vinylpyridinium Compounds. III. On the Mechanism of the 'Spontaneous Polymerization' of 4-Vinylpyridinium Salts in Water," *Makromol. Chem.*, 142 (1971), 319.

5. Kabanov, V.A., Patrikeeva, T.I., Kargina, O.V., and Kargin, V.A., "Organized Polymerization of Vinylpyridinium Salts," *J. Polym. Sci., Part C*, no. 23, pt. 1 (1968), 357.

6. Kargina, O.V., Kabanov, V.A., and Kargin, V.A., "Polymerization of 4-Vinylpyridine on Polyacid Matrices," *J. Polym. Sci., Part C*, no. 22, pt. 1 (1968), 339.

7. Kabanov, V.A., Petrovskaya, V.A., and Kargin, V.A., "Kinetics and Mechanism of Polymerization of 4-Vinylpyridine on Macromolecules of Polyacrylic and Poly-L-Glutamic Acids," *Polym. Sci. USSR*, 10, no. 4 (1968), 1077.

8. Chapiro, A., Bex, G., Jendrychowska-Bonamour, A.M., and O'Neill, T., "Preparation of Permselective Membranes by Radiation Grafting of Hydrophilic Monomers into Polytetrafluoroethylene Films," in *Addition and Condensation Polymerization Processes*, no. 91, *Advances in Chemistry Series*, R.F. Gould, ed., Washington, D.C.: American Chemical Society (1969), 560.

9. Chapiro, A., "Controlled Propagation in Associated Monomer Aggregates," *Pure Appl. Chem.*, 30, pt. 1 (1972), 77.

10. Ferguson, J. and Shah, S.A.O., "Polymerization in an Interacting Polymer System," *Eur. Polym. J.*, 4, no. 3 (1968), 343.

11. Sakaguchi, Y., Tamaki, K., Shimizu, T., and Nishino, J., "Stereoregularities of Vinyl Polymers Obtained by Radical Polymerization in the Presence of Polyvinylpyrrolidone," *Kobunshi Kagaku*, 27 (1970), 284.

12. Sakaguchi, Y., Harimoto, Y., Tamaki, K., and Nishino, J., "Stereoregularity of Poly(methylacrylate), Poly(acrylic acid) and Poly(vinyl acetate) Obtained by Radical Polymerization in the Presence of Vinyl Polymers," *Kobunshi Kagaku*, 28 (1971), 653.

13. Endo, T., Numazawa, H., and Okawara, M., "Syntheses and Reactions of Functional Polymers. LVIII. Polymerization of N-vinyl-2-oxazolidone in the Presence of Polymethacrylic Acid," *Makromol. Chem.,* 146 (1971), 247.

14. Ferguson, J. and Shah, S.A.O., "Further Studies on Polymerizations in Interacting Polymer Systems," *Eur. Polym. J.,* 4, no. 5 (1968), 611.

15. Bamford, C.H. and Shiiki, Z., "Free-radical Template Polymerization," *Polymer,* 9, no. 11 (1968), 595.

16. Miyamoto, T. and Inagaki, H., "The Stereocomplex Formation in Poly(methyl methacrylate) and the Stereospecific Polymerization of its Monomer," *Polym. J.,* 1, no. 1 (1970), 46.

17. Buter, R., Tan. Y.Y., and Challa, G., "Radical Polymerization of Methyl Methacrylate in the Presence of Isostatic Poly(methyl methacrylate)," *J. Polym. Sci., Part A-1,* 10, no. 4 (1972), 1031.

18. Shima, K., Kakui, Y., Kinoshita, M., and Imoto, M., "Vinyl Polymerization. 277. Polymerization of Vinylpyridines in the Presence of Poly(maleic anhydride)," *Makromol. Chem.,* 154 (1972), 247.

19. Kakui, Y., Shima, K., Kinoshita, M., and Imoto, M., "Vinyl Polymerization. 282. Copolymerization of 4-Vinylpyridine and p-Chlorostyrene in the Presence of Poly(maleic anhydride)," *Makromol. Chem.,* 155 (1972), 299.

20. Gaylord, N.G., "Donor–Acceptor Complexes in Copolymerization. VI. A Proposed Mechanism for the Homopolymerization of Comonomer Complexes to Alternating Copolymers," *J. Polym. Sci., Part C,* no. 31 (1970), 247.

21. Gaylord, N.G., "Donor–Acceptor Complexes in Copolymerization. XXVIII. Role of Matrices in Polymerization of Comonomer Charge Transfer Complexes," *J. Macromol. Sci., Chem.,* A6, no. 2 (1972), 259.

22. Gaylord, N.G., "Donor–Acceptor Complexes in Copolymerization. XXXI. Alternating Copolymer Graft Copolymers. VI. Syntheses in the Absence and in the Presence of Complexing Agents, in *Polymerization and Polycondensation Processes,* R.F. Gould, ed., Washington, D.C.: American Chemical Society (1973), in press.

23. Gaylord, N.G., Patnaik, B.K., and Stolka, M., "Donor–Acceptor Complexes in Copolymerization. XXIX. Alternating Copolymer Graft Copolymers. IV. Grafting of Alternating Copolymers Onto Precomplexed Polymers," *J. Polym. Sci., Part B,* 9 (1971), 923.

24. Gaylord, N.G. and Anand, L.C., "Donor–Acceptor Complexes in Copolymerization. XXIV. Alternating Copolymer Graft Copolymers. III. Cellulose Graft Copolymers. I. Grafting of Alternate Styrene–Acrylonitrile Copolymers onto Cellulose in Presence of Zinc Chloride," *J. Polym. Sci., Part B.,* 9 (1971), 617.

25. Gaylord, N.G.; "A Proposed New Mechanism for Catalyzed and Uncatalyzed Graft Polymerization Onto Cellulose," *J. Polym. Sci., Part C,* no. 37 (1972), 153.

26. Imoto, M., Kondo, M., and Takemoto, K., "Vinyl Polymerization. 107. Grafting of Vinyl Monomers on Different Fibers," *Makromol. Chem.,* 89 (1965), 165.

27. Imoto, M., Takemoto, K., and Kondo, M., "Vinyl Polymerization. 143. Radical Polymerization of Methyl Methacrylate on Cellulose and Silk in the Absence of Initiators," *Makromol. Chem.,* 97 (1966), 74.

28. Imoto, M., Takemoto, K., Okura, A., and Izubayashi, M., "Vinyl Polymerization. 194. Polymerization of Methyl Methacrylate in the Presence of Partially Hydrolyzed Poly(vinyl acetate) and Water," *Makromol. Chem.,* 113 (1968), 111.

29. Imoto, M., Takemoto, K., and Sutoh, H., "Vinyl Polymerization in the Presence of Macromolecules, Water and Carbon Tetrachloride: A Macromolecule as a Part of a Complex Initiator," *Makromol. Chem.,* 110 (1967), 31.

30. Imoto, M., Iki, Y., Kinoshita, M., and Takemoto, K., "Vinyl Polymerization. 232. On the Initiating Species of the Polymerization of Methyl Methacrylate Initiated with Cellulose and Water. Accelerating Effect of Carbon Tetrachloride on the Polymerization," *Makromol. Chem.,* 122 (1969), 287.

31. Imoto, K., Iki, Y., Kawabata, Y., and Kinoshita, M., "Vinyl Polymerization. 260. Contribution of a Slight Amount of Cupric Ion to the Polymerization of Methyl Methacrylate Initiated with Cellulose and Water," *Makromol. Chem.,* 140 (1970), 281.

32. Gaylord, N.G. and Anand, L.C., "Cellulose Graft Copolymers. II. Grafting of Methacrylonitrile in the Presence of Ceric Ion and Ammonium Persulfate," *J. Polym. Sci., Part B,* 10 (1972), 285.

33. Sakaguchi, Y., Mimuro, K., Tamaki, K., and Nishino, J., "Stereoregularities of Some Vinyl Polymers Graft Polymerized Onto Rayon and Poly(vinyl alcohol) Fibers," *Kobunshi Kagaku,* 26 (1969), 787.

34. Nara, S. and Matsuyama, K., "Methyl Methacrylate-co-Butadiene Grafted Cellulose. I. Synthesis and Characterization," *Kobunshi Kagaku,* 25 (1968), 840.

35. Rebek, M., Schurz, J., Stoeger, W., and Popp, W., "Graft Polymerization of Cellulose with Two Vinyl Monomers," *Monatsh. Chem.,* 100 (1969), 532.

36. Sakurada, I., Okada, T., Hatakeyama, S., and Kimura, F., "Radiation Induced Graft Copolymerization Onto Cellulose and Poly(vinyl alcohol) Fibers with Binary Mixtures of Comonomers," *J. Polym. Sci., Part C,* no. 4 (1963), 1233.

37. Gaylord, N.G. and Anand, L.C., "Donor–Acceptor Complexes in Copolymerization. XXXIV. Alternating Copolymer Graft Copolymers. VII. Cellulose Graft Copolymers. III. Grafting of Alternating Styrene–Acrylonitrile and Styrene–Methacrylonitrile Copolymers Onto Cellulose in Absence of Metal Halide," *J. Polym. Sci., Part B,* 10 (1972), 305.

38. Dilli, S. and Garnett, J.L., "A Charge-Transfer Theory for the Interpretation of Radiation-Induced Grafting of Monomers to Cellulose," *J. Polym. Sci., Part A-1,* 4 (1966), 2323.

39. Dilli, S. and Garnett, J.L., "Radiation-Induced Reactions with Cellulose. II. The Trommsdorff Effect: Dose and Dose Rate Studies Involving Copolymerization with Monomers in Methanol," *J. Appl. Polym. Sci.,* 11 (1967), 839.

40. Dilli, S. and Garnett, J.L., "Radiation-Induced Reactions with Cellulose. III. Kinetics of Styrene Copolymerization in Methanol," *J. Appl. Polym. Sci.,* 11 (1967), 859.

41. Siegel, S.M., "The Biosynthesis of Lignin: Evidence for the Participation of Celluloses as Sites for Oxidative Polymerization of Eugenol," *J. Amer. Chem. Soc.,* 78 (1956), 1753.

42. Siegel, S.M., "Non-enzymic Macromolecules as Matrices in Biological Synthesis: The Role of Polysaccharides in Peroxidase-Catalyzed Lignin Polymer Formation from Eugenol," *J. Amer. Chem. Soc.,* 79 (1958), 1628.

43. Imoto, M., Takemoto, K., Azuma, H., Kita, N., and Kondo, M., "Vinyl Polymerization. 170. Grafting of Some Acrylates and Methacrylates Onto Cellulose and Silk in Absence of Radial Initiators," *Makromol. Chem.,* 107 (1967), 188.

44. Takemoto, K., Kondo, M., Iwasaka, T., and Imoto, M., "Vinyl Polymerization. 195. Effect of Trace Metals on the Polymerization of Methyl Methacrylate in the Presence of Macromolecules and Water: Activation Analysis of Metals in Cellulose and Silk," *Makromol. Chem.,* 112 (1968), 110.

45. Arai, K., Negishi, M., and Okabe, T., "Infrared Spectroscopy of Graft Polymers Separated from Graft Copolymers of Wool and Silk with Methyl Methacrylate," *J. Appl. Polym. Sci.,* 12 (1968), 2585.

46. Tanaka, Z., "Graft Polymerization of Vinyl Monomers Onto Various Fibers and Cloths in the Absence of Initiators. V. Polymerization of Methyl Methacrylate in the Presence of Wool Fibers," *Kogyo Kagaku Zasshi,* 74 (1971), 1683.

47. Kantouch, A., Hebeish, A., and Bendak, A., "Ce^{IV} Initiated Graft Polymerization of Methyl Methacrylate on Wool Fibers," *Eur. Polym. J.,* 7 (1971), 153.

48. Okamoto, K. and Yamamoto, T., "Polymerization of Methyl Methacrylate in the Presence of Collagen Fiber," *Kogyo Kagaku Zasshi,* 74 (1971), 527.

49. Hayashi, S. and Imoto, M., "Vinyl Polymerization. 203. Polymerization of Methyl Methacrylate in the Presence of Nylon-6," *Kogyo Kagaku Zasshi,* 71 (1968), 577.

50. Hayashi, S. and Imoto, M., "Vinyl Polymerization. 218. Polymerization of Methyl Methacrylate in the Presence of Nylon-6 Fiber and Water," *Angew. Makromol. Chem.,* 6 (1969), 46.

51. Imoto, M., Tanaka, A., Ueno, K., and Takemoto, K., "Vinyl Polymerization. 265. Polymerization of Methyl Methacrylate Initiated with Nylon-6 and Water in the Absence or Presence of Carbon Tetrachloride. Contribution of a Trace Amount of Cupric Ion to the Polymerization," *Angew. Makromol. Chem.,* 18 (1971), 55.

52. Ueno, K., Kinoshita, M., and Imoto, M., "Vinyl Polymerization. 273. Polymerization of Methyl Methacrylate Initiated with 6-Nylon and Water in the Presence of a Trace Amount of Cupric Ion," *Kogyo Kagaku Zasshi,* 74 (1971), 1921.

53. Sakaguchi, Y., Nishino, J., Tamaki, K., and Motomiya, H., "Stereoregularities of Some Vinyl Polymers Graft-Polymerized Onto Nylon Fiber," *Kobunshi Kagaku,* 25 (1968), 756.

54. Sakurada, I., Sakaguchi, Y., Omura, Y., Shikata, E., and Tone, M., "Stereoregularity of Poly(vinyl acetate) Grafted Onto Nylon Fiber," *Kobunshi Kagaku,* 26 (1969), 794.

55. Imoto, M., Takemoto, K., and Sutoh, H., "Polymerization of 1-Vinyl-2-Methylimidazole Initiated by Carbon Tetrachloride, Water and Cellulose," *Bull. Chem. Soc. Jap.,* 40 (1967), 413.

56. Imoto, M., Takemoto, K., Sutoh, H., and Azuma, H., "Participation of Imidazole in Vinyl Polymerization in the Absence of Radical Initiator," *J. Polym. Sci., Part C,* no. 22 (1968), 89.

57. Imoto, M., Takemoto, K., Otsuki, T., Ueda, N., Tahara, S., and Azuma, H., "Vinyl Polymerization. 177. Polymerization of Acrylonitrile Initiated with the Imidazole–Carbon Tetrachloride System," *Makromol. Chem.,* 110 (1967), 37.

3. Synthesis of Block Copolymers via Anionic Polymerization

HENRY L. HSIEH

Phillips Petroleum Company
Bartlesville, Oklahoma

ABSTRACT

Anionic polymerization initiated with soluble organoalkali metal initiators enables the preparation of block copolymers in which the sequence and length of the blocks are controlled by incremental addition of monomers. Different types of block copolymers are illustrated, and the general techniques that can be used for their synthesis are discussed.

Anionic polymerization initiated with soluble organoalkali metal initiators is a homogeneous and, through the judicious choice of experimental conditions, a termination-free and transfer-free system. It allows the preparation of polymers with extremely narrow molecular-weight distribution [1], approaching the Poisson distribution as predicted by Flory, and block copolymers in which the sequence and length of the block are controlled by incremental (or sequential) addition of monomers. This general method of preparing block copolymers is readily adaptable to commercial production. Indeed, a number of block copolymers are manufactured this way.

The two key factors for accomplishing successful synthesis of well-defined block copolymers by this general method are that all the chains are initiated nearly simultaneously and that the system is free from premature termination and transfer reactions. These can be achieved by selection of the proper initiator and by rigorous purification of solvent and monomer. Among the alkali metal compounds, organolithium compounds are particularly suited for this type of synthesis. The requirement that all polymer chains start at the same time depends entirely on the relative rates of initiation and propagation. A list of alkyllithium initiators in the order of their relative reactivities toward diene and styrene monomers, as well as the solvent effect on initiation rates is shown in Table I. The large differences among these hydrocarbon-soluble, organolithium initiators can be best illustrated by citing some semiquantitative data. In cyclo-

TABLE I

Relative Reactivities of Organolithiums in Initiation [1,2,3]

For dienes:
sec-BuLi > i-PrLi > t-BuLi > i-BuLi > n-BuLi and EtLi
For styrenes:
menthyl Li > sec-BuLi > i-PrLi > i-BuLi > n-BuLi and EtLi > t-BuLi
Solvent effect:
toluene > benzene ≫ n-hexane > cyclohexane

hexane and with styrene as monomer, the rate of initiation for menthyllithium is as much as fifty times greater than that for sec-butyllithium, which, in turn, is approximately sixty times greater than that for n-butyllithium. Even slower re-acting toward styrene than n-butyllithium is t-butyllithium. The initiation reactions are quite a bit faster in aromatic solvents, as much as thirty to forty times faster than in aliphatic solvents. The propagation reactions are also faster in aromatic solvent, but not to the same extent as the increase in initiation reaction, and the use of aromatic solvent is therefore more desirable in this regard. However, toluene must be used with care since this solvent is known to act as a chain-transfer agent in polymerization systems involving sodium and lithium counterions [4,5,6]. Introduction of very small amounts of polar solvents, such as tetrahydrofuran or triethylamine, generally increases the initiation rates considerably more than the increase in the chain-growth rates [7]. The amounts required are not enough to affect the stereochemistry of diene polymerization.

The simplest block copolymers are the A–B type, which can be conveniently prepared with a monometal initiator, such as butyllithium, and addition of monomer B after monomer A is completely polymerized. Vinyl monomers containing functional groups such as acrylates, or polar monomers such as ε-caprolactone or ethylene sulfide may terminate the "living" end or may not initiate polymerization of some second monomer. However, these monomers can be used as monomer B to form the A–B type of block copolymer, in some cases by adding and polymerizing monomer B at very low temperatures to minimize side reactions with the functional group.

One variation of the A–B type is the A/B–B type. In this case, mixtures of A and B are polymerized randomly first, and then the monomer B is added. A closely similar type can also be prepared from a monomer mixture discussed later.

The A–B–A block copolymers can be prepared in a number of ways. One can use a monometal initiator and add monomers in three increments. One can use a monometal initiator to form an A–B block copolymer first and then couple the living end with an active difunctional compound such as an active dihalide or diepoxide. Carbon dioxide and phosgene also can be successfully used as coupling agents. The advantage of this method is that it requires only two increments of monomer addition, which minimizes termination due to residual impurities in monomer. However, it demands high precision in the stoichiometry of the coupling reaction and its efficiency. The third method involves the use of an initiator that will yield polymer anions propagating at two ends, so that each increment of monomer after the first forms two blocks. In this case the middle block forms first, which limits the types of monomer one can use for the middle block. Initiators such as dilithium compounds prepared from the reactions of lithium metal with dimethylbutadiene, dichlorobutane or naphthalene, and sodium or potassium complexes of naphthalene or bisphenyl are often used. The latter group yields dianions via an electron-transfer process.

The A-B-C block polymers have three chemically dissimilar segments joined end-to-end. Here again monomers such as acrylonitrile, methyl methacrylate, caprolactone, and ethylene sulfide, if used, must be used in the last increment.

Block polymers of the (A-B)n type, as the block lengths become smaller and smaller, become more nearly a kind of alternating copolymer. (A-B)n types can be prepared by incremental addition of monomers alternating between A and B again and again, using either a monometal or dimetal initiator. However, trace impurities in monomers that will terminate the active chain-end are the limiting factor. A better way to make this type of block polymer is first to make a relatively low-molecular-weight A-B-A type having two active ends and then to couple this living block copolymer with a difunctional compound. In the laboratory, with proper care and selection of conditions and coupling agent, the n can be as high as twenty by the latter method.

By applying this technique of the sequential addition of monomer, block copolymers having rubbery and glassy blocks, crystalline and amorphous blocks, rigid and flexible blocks, polar and nonpolar blocks of widely different T_g values are prepared. Table II shows some of the combinations known to form block copolymers via anionic polymerization.

Commercially produced thermoplastic elastomers, A-B-A block copolymers are based on styrene (A) and butadiene or isoprene (B), and they exhibit excellent tensile strengths at room temperature. However, tensile strength of these

TABLE II

Block Copolymers Prepared with Anionic Initiators

Monomer I	Monomer II
Isoprene, butadiene	styrene, α-methylstyrene
Diene, styrene	4-vinylbiphenyl, vinylnaphthalene, vinylpyrene
Diene, styrene	2-vinylpyridine
Diene, styrene	ethylene sulfide, propylene sulfide
Diene, styrene	formaldehyde
Diene, styrene	ethylene oxide
Diene, styrene	acrylates, acrylonitrile
Diene, styrene	octamethylcyclotetrasiloxane, dimethylsiloxane
Diene, styrene	n-butylisocyanate
Diene, styrene	ε-caprolactone
Propylene sulfide	ethylene sulfide
2-methylthiocyclobutane	styrene, ethylene sulfide
Diene	styrene/1,1-diphenylethylene (1/1)
Butadiene	1,1-bis(trifluoromethyl)-1,3-butadiene

polymers drops sharply as the temperature is increased. This has led many workers to search for a replacement for styrene as the monomer that gives glassy endblocks having a higher T_g than exhibited by polystyrene. α-methylstyrene, which can be polymerized with a lithium system in a polar solvent at a low temperature (α-methylstyrene has a ceiling temperature of ~6°C for a 1 M solution), was successfully used for making the A-B-A polymer (butadiene or isoprene as monomer B). It showed improvement of the tensile strength at an elevated temperature [8]. Monomers such as vinylbiphenyl, vinylnaphthalene, and vinylpyrene were tried for the same reason. Styrene and 1,1-diphenylethylene copolymerize, when an organolithium initiator is used, to an alternating copolymer with a softening point of about 182°C. When this alternating copolymer replaced polystyrene, the product again showed better tensile strength at 70°C [9].

Another interesting A-B-A-type block copolymer is poly(ethylene sulfide)–poly(propylene sulfide)–poly(ethylene sulfide). These two monomers are polymerized via thiolate lithium active ends [10,11,12]. Poly(ethylene sulfide) is crystalline, and poly(propylene sulfide) is amorphous and rubbery. Unlike the episulfides, the four-membered ring sulfide compound 2-methylthiocyclobutane reacts with either alkyllithium or polymer-lithium to form a primary carbanion species; therefore, after its polymerization, a hydrocarbon monomer such as styrene can be attached and subsequently polymerized. One can also add ethylene sulfide to the dilithium-initiated poly-2-methylthiobutane to form an A-B-A-type block copolymer. Poly-2-methylthiobutane ($T_g = -59$°C) is colorless, amorphous, and elastomeric; while poly(ethylene sulfide) is highly crystalline.

There is yet another class of block copolymer that can be either A-B, A-B-A, or even (A-B)n type. Diene can be polymerized with an organolithium initiator to give different microstructures, depending upon the presence or absence of polar compounds such as ethers. Therefore, one can use a single monomer such as butadiene and, by adjusting polymerization conditions, produce an essentially steroblock copolymer.

In an indirect way one can prepare a diblock, triblock, or multiblock copolymer of polystyrene and polyolefin by hydrogenating the polydiene block. This type of polymer should have improved environmental resistance compared to the unsaturated version from which it is derived. Thermoelastomeric as well as thermoplastic products with very good physical properties have been made this way [13].

Under certain conditions some monomer mixtures anionically polymerize directly to block copolymers. For example, in a hydrocarbon solution of butyllithium, styrene polymerizes more rapidly than does butadiene. However, when a mixture of butadiene and styrene is polymerized, the "living" polymer anions are rich in butadiene until late in the reaction, whereafter styrene suddenly increases. Figure 1 illustrates the comparison in terms of conversion and time of reaction. The inflection point in the conversion curves for the mixture can be easily identified by the appearance of an orange-yellow color (styryl anion) in the solution. For the purpose of this discussion, the butadiene-to-styrene weight ratio is 75 to 25. As shown in Figure 2, the type of hydrocarbon solvent changes the overall rate but not the general shape of the curves. As one analyzes samples at various conversions, the results shown in Figure 3 are obtained. Styrene contents are initially lower than in the monomer charge, gradually increase until the inflection point of the conversion curve in Figure 2 is attained, and thereafter increase very rapidly. Furthermore, when these same samples are analyzed by oxidation degradation, polystyrene segments are recovered only after the inflection points are reached and thereafter increase. This phenomenon is independent of the polymerization temperature (Figure 4). Since the final polymers are essentially homogeneous in composition and molecular weight, it follows that the process has resulted in a diblock copolymer, one segment being a butadiene-styrene copolymer rich in butadiene, the other a polystyrene block. This structure is type A/B-B. On the other hand, as a result of the counterion being changed from lithium to potassium, styrene was almost completely consumed before the conversion had reached the halfway point (Figure 5). In this case, one is making a block copolymer of A/B-A type, just the reversal of the A/B-B type initiated with alkyllithium [14]. Other vinylaromatic monomers can also be used with diene to form block copolymers without the application of incremental addition of monomers. It is well known by now, of course, that in the presence of an ether or tertiary amine, or certain sodium or potassium compounds, the monomer mixtures initiated with alkyllithium compound result in a random copolymer.

The anomolous copolymerization behavior of the diene–styrene system in hydrocarbon solvent with an alkyllithium initiator can be explained quite satisfactorily by copolymerization kinetics [1]. The cross-propagation rate of styryl anion to dienyl anion is extremely fast, but the opposite is true for the dienyl anion to styryl anion ($k_{SD} \gg k_{SS} > k_{DD} > k_{DS}$):

$$D^- + S \xrightarrow{k_{DS}} S^-$$

$$D^- + D \xrightarrow{k_{DD}} D^-$$

$$S^- + D \xrightarrow{k_{SD}} D^-$$

$$S^- + S \xrightarrow{k_{SS}} S^-$$

Table III shows a few selective reactivity ratios in lithium-initiated copolymerizations. It should be pointed out that although kinetics explain the "inversion" behavior in copolymerization of diene–styrene quite adequately, it is nevertheless only a formal account of what happens. It does not provide a mechanism that is responsible for this behavior.

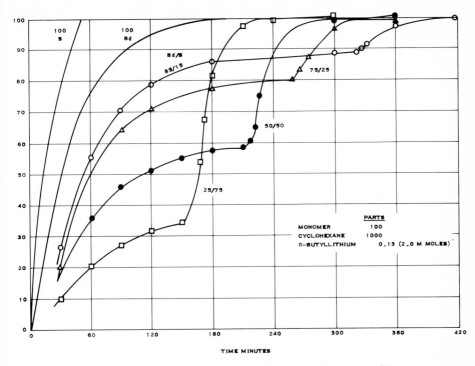

Figure 1. Polymerization of butadiene–styrene in cyclohexane at 50°C.

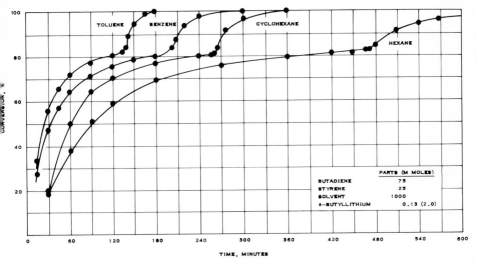

Figure 2. Polymerization of butadiene–styrene in different solvents at 50°C.

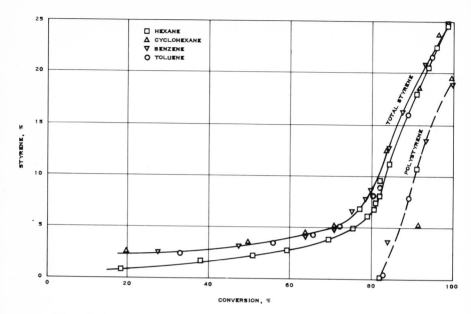

Figure 3. Copolymerization of styrene from butadiene–styrene (75/25) at 50°C.

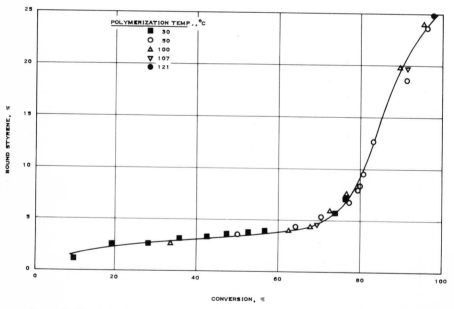

Figure 4. Styrene incorporation from butadiene–styrene (75/25) in cyclohexane solutions at 30–121°C.

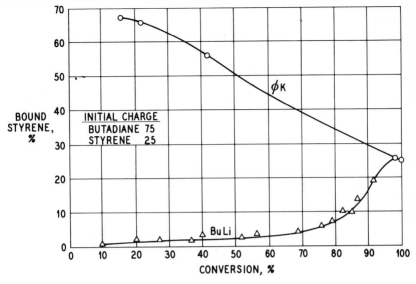

Figure 5. Styrene incorporation.

TABLE III

Reactivity Ratios in Lithium-Initiated Copolymerization [1]

M_1	M_2	Solvent	Temperature °C	r_1^*	$r_2^\dagger$
Styrene	butadiene	benzene	30–50	0.05	15
		toluene	25	0.1	12.5
		cyclohexane	40	<0.04	26
		THF	30	0.77	1.03
Styrene	isoprene	benzene	30	0.14	7
		toluene	27	0.25	9.5
		cyclohexane	40	0.05	16.6
Styrene	1,1-diphenyl-ethylene	toluene	30	0.04	$-\ddagger$

$^*r_1 = k_{11}/k_{12}$.
$^\dagger r_2 = k_{22}/k_{21}$.
$^\ddagger k_{22} = 0$.

Summary

The unique features of homogeneous anionic polymerization allow the preparation of well-defined block copolymers as well as random copolymers. Many of the block copolymers possess unusual properties that are interesting from both theoretical and practical points of view.

References

1. Hsieh, H.L. and Glaze, W.H., "Kinetics of Alkyllithium Initiated Polymerizations," *Rubber Chem. Tech.*, 43, no. 1 (1970), 22.

2. Hsieh, H.L., "Kinetics of Polymerization of Butadiene, Isoprene and Styrene with Alkyllithiums. Part II. Rate of Initiation," *J. Polym. Sci., Part A*, 3, no. 1 (1965), 163.

3. Selman, C.M. and Hsieh, H.L., "Effect of Aggregate Size on Alkyllithium Initiated Polymerizations," *J. Polym. Sci., Part B*, 9 (1971), 219.

4. Robertson, R.E. and Marion, L., "The Mechanism of Sodium-Initiated Polymerization of Diolefins," *Can. J. Res.*, 26B (1948), 657.

5. Gatzke, R.L., "Chain Transfer in Anionic Polymerization," *J. Polym. Sci., Part A-1*, 7 (1969), 2281.

6. Brooks, B.W., "Chain Transfer in the Anionic Polymerization of Styrene," *Chem. Commun.* (1967), 68.

7. U.S. Patent No. 3,639,521, February 1, 1972, "Polar Compound Adjuvants for Improved Block Polymers Prepared with Primary Hydro-carbyllithium Initiators," H.L. Hsieh to Phillips Petroleum Company, Bartlesville, Oklahoma.

8. Fetters, L.J. and Morton, M., "Synthesis and Properties of Block Polymers, I. Poly(α-methylstyrene)-Polyisoprene-Poly(α-methylstyrene)," *Macromolecules*, 2, (1969), 453.

9. Trepka, W.J., "Synthesis and Properties of Block Polymers of 1,1-Diphenylethylene/Styrene and Butadiene," *J. Polym. Sci., Part B*, 8 (1970) 499.

10. Morton, M. and Kammereck, R.F., "Nucleophilic Substitution at Bivalent Sulfur. Reaction of Alkyllithium with Cyclic Sulfides," *J. Amer. Chem. Soc.*, 92 (1970), 3217.

11. Morton, M., Kammereck, R.F., and Fetters, L.J., "Synthesis and Properties of Block Polymers. II. Poly(α-methylstyrene)-Poly(propylene sulfide)-Poly(α-methylstyrene)," *Macromolecules*, 4 (1971), 11.

12. Morton, M.M. and Mikesell, S.L., "ABA Block Copolymers of Dienes and Cyclic Sulfides, *Polym. Prepr., Amer. Chem. Soc., Div. Polym. Chem.*, 13, no. 1 (1972), 61.

13. Pendleton, J.F., Hoeg, D.F., and Goldberg, E.P., "Novel Heat Resistant Plastics from Hydrogenation of Styrene Polymers," *Polym. Prepr., Amer. Chem. Soc., Div. Polym. Chem.*, 13, no. 1 (1972), 427.

14. Wafford, C.F. and Hsieh, H.L., "Copolymerization of Butadiene and Styrene by Initiation with Alkyllithium and Alkali Metal *tert*-Butoxides," *J. Polym. Sci., Part A-1*, 7 (1969), 461.

4. Synthesis of Siloxane–Organic Block Polymers with Organometallic Initiators

PETER C. JULIANO, DANIEL E. FLORYAN,* ROBERT W. HAND,†
AND DOUGLAS D. KARTTUNEN‡

General Electric Research and Development Center
Schenectady, New York

ABSTRACT

The synthesis of siloxane–organic block polymers with organometallic initiators is discussed in detail. The effects of initiator structure and metal counterion, of solvent polarity, of Lewis base promoters, and of cyclosiloxane monomer type on siloxane polymerizations with respect to block polymer formation are also reviewed.

In a more specific example, the synthesis of acrylic–siloxane block polymers with ambident initiators via homogeneous anionic polymerization techniques is described. This class of initiators possesses two anionic sites of unequal reactivity; one is selective for the siloxane polymerization, the other for the acrylic polymerization. The variables controlling the reproducible synthesis of block polymers are outlined in detail. Some of these variables are the temperature of the siloxane polymerization, the molecular-weight distribution of the poly(methyl methacrylate) block, the total molecular weight, and the composition and composition distribution of the block polymers. The composition distribution of the block polymers is verified by foam fractionation and analysis of the components. Polydimethylsiloxane domains in a poly(methyl methacrylate) matrix are observed by electron microscopy and are less than 400 Å, thus accounting for the optical clarity of these block polymers.

Polymerization of cyclopolysiloxanes with alkali-metal hydroxides was reported first by Hyde in 1949 [1,2]. Subsequent studies elucidated the kinetics [3–6], mechanism [7–11], and effect of Lewis bases as promoters [12–15],

*Plastics Business Department, Central Research, General Electric Company, Pittsfield, Massachusetts.

†University of California, Berkeley.

‡Michigan Technological University, Houghton.

all of which is summarized in several reviews [16,17]. The alkali-metal siloxano-late species capable of polymerizing cyclopolysiloxanes are not sufficiently basic to initiate the polymerization of vinyl or diene monomers. Thus, organic-siloxane block polymers must be synthesized by first polymerizing the vinyl or diene monomers [18-21] under termination-free conditions [22]. Lithium-derived initiators [23-30] provided an additional control over the polymeriza-tion of cyclopolysiloxanes, and narrow-molecular-weight-distribution poly-mers [23,24] of predictable molecular weights were obtained free of higher cyclics. In addition, with these initiators all-siloxane [23,31,32] and organic-siloxane block polymers [33] were synthesized free of contaminating polysilox-ane homopolymer and cyclics. The synthesis [33-45] and properties [46-48] of well-defined styrene–siloxane [33-43] and α-methyl-styrene–siloxane [44,45] block polymers have been reported, and various applications [49-55] have been defined.

Recently the synthesis of acrylic–siloxane block polymers [56] with ambient initiators [57-62] has been described. This method, detailed below, differs from the conventional approach [33,56] in which the methyl methacrylate block was polymerized first with organolithium initiators [63-67]. The lithium methoxide produced during both the initiation and propagation reactions [68] of the methyl methacrylate polymerization will polymerize the siloxane monomer, hexamethylcyclotrisiloxane. Thus, block polymers produced with the conven-tional method showed low incorporation of siloxane and produced cloudy films when cast from chlorinated aliphatic solvents [56]. The block polymers pro-duced with the ambient initiator approach showed a much greater siloxane in-corporation (for the same feed composition) and produced clear films when cast from chlorinated aliphatic solvents [56].

Experimental

Materials

1. *Monomers.* *a.* Hexamethylcyclotrisiloxane (D_3) was obtained from the Silicone Products Business Department of the General Electric Company, Water-ford, New York. Prior to use, the D_3 was refluxed for three hours over calcium hydride (CaH_2), then fractionally distilled at atmospheric pressure (bp 134°C) [69] under a dry nitrogen atmosphere. Purified D_3 was stored in a dry-nitrogen-filled dry box, as were all other monomers, initiators, and solvents.

b. *Cis*-2,4,6-trimethyl-2,4,6-triphenylcyclotrisiloxane (CMPT) was prepared and purified according to the method of Young et al. [70]. The CMPT, ana-lyzed by NMR, showed only one silicon methyl resonance at 9.46 τ [71] (THF, cyclohexane internal standard).

c. Methyl methacrylate (MMA), MC&B Chromatoquality, was stirred over CaH$_2$ for 24 hours at room temperature, then placed over fresh CaH$_2$ prior to distillation. The distillation was carried out under a N$_2$ atmosphere, and a middle-fraction boiling at 100°C was collected. MMA was purified for each experiment just prior to use.

2. *Initiators.* *a.* Dilithiobenzophenone dianion (BPDA) [56,59] was prepared from pure, hexane-recrystallized benzophenone (1.82 g, 10 mmol) and freshly scraped lithium wire (0.35 g, 50 mg-atoms) in 50 ml of dry, freshly distilled tetrahydrofuran (THF) in an argon-filled dry bag. This mixture was contained in a 2-oz screw-capped bottle sealed with a conical, polyethylene-lined cap. The cap was securely taped and the bottle was shaken for 48 hours at 25°C to insure complete formation of the BPDA initiator.

b. 1,1-diphenylhexyllithium (DPHL) [72,73] was prepared by the addition of 1.25 ml of *n*-butyllithium in hexane (2.50 mmol) to 0.45 g (2.50 mmol) of 1,1-diphenylethylene in 50 ml of dry THF at −30°C. The characteristic dark red-orange color of the 1,1-diphenylmethyl carbanion appeared immediately. After one-half hour at −30°C, the solution was allowed to warm to room temperature in a dry-nitrogen-filled dry box.

3. *Solvents.* *a.* Tetrahydrofuran (THF) was dried over CaH$_2$. Prior to use, the THF was fractionally distilled from a sodium–biphenyl complex under a N$_2$ atmosphere into a clean, dry glass bottle and transferred to a N$_2$ dry box. THF was purified and dried for each experiment.

b. Hexamethylphosphortriamide (HMPA) was distilled from CaH$_2$ on a spinning band column. A middle fraction, bp 71°C/1 mm, was collected and transferred to a N$_2$ dry box.

Polymerizations

1. *Block polymers.* This procedure is described in the section below on block polymer synthesis and characterization as the "modified synthesis" of block polymers. It is typical of that of all the block polymers prepared, and variations of initiator levels and of the relative amounts of the MMA and D$_3$ provide different total molecular weights and compositions, respectively.

To a 16-oz screw-capped glass bottle in a dry N$_2$-filled dry box were charged 150 ml of dry, distilled THF and 10 g (45.1 mmol) of D$_3$. The bottle was placed in a cooling bath, and the temperature was lowered to 0°C. The BPDA-initiator in THF was added (2.5 ml, 0.50 mmol). Within 30–45 minutes the color of the solution changed from red-purple to orange-brown. To this solution was added 0.21 ml of HMPA (1.0 mmol), and the polymerization was continued for 8 hours at 0°C. The bath temperature was lowered to −45°C, and upon equilibration of the flask contents at this temperature, 20 ml (18.8 g, 188 mmol) of freshly distilled MMA were added, with vigorous mechanical stirring. After 2 hours at

-45°C, the polymerization was terminated with acetic acid in THF while the reaction mixture was still cold. The room-temperature polymer solution was precipitated into 2 liters of methanol, filtered, and dried in a vacuum oven at 80°C. A white fibrous polymer was obtained (29.8 g, 99.3 percent), with an intrinsic viscosity (IV) of 0.61 dl/g in chloroform and a dimethylsiloxane (D) content of 35.2 percent by weight.

A sample of this polymer (10.0 g) was extracted with hexane for 48 hours with a Soxhlet extraction apparatus to yield 8.77 g of hexane insoluble (HI) polymer, 87.7 percent HI. The extracted polymer had an IV of 0.66 dl/g and a D content of 21.6 percent by weight.

2. *Poly(methyl methacrylate) (PMMA)*. To a clean, dry 16-oz screw-capped bottle were added 100 cc of THF and 5.0×10^{-4} mol of freshly prepared DPHL. The solution was cooled to -40°C. When the effect of HMPA was being examined, it was added at this point, 1:1 based on the DPHL. Freshly distilled MMA, 9.4 g, was added via syringe, with rapid stirring. After 1 hour at -40°C, the polymerization was terminated with several drops of glacial acetic acid, HOAc. The polymer solution was warmed to room temperature and then precipitated into a tenfold excess of methanol. The polymer was isolated by filtering directly into a tared fritted-glass funnel. Yields were measured after overnight drying of the polymer in a vacuum oven at 80°C.

3. *CMPT polymers*. To a clean, dry 4-oz screw-capped bottle were added 2.00 g CMPT and 50 ml THF. The solution was cooled to 0°C in a N_2-filled dry box, and 0.5 cc of BPDA (0.5 mmol) in THF were added. When the effect of HMPA was being examined, it was added (1:1 based on Li) after the initiation was complete (~45 minutes). The polymerization was terminated with glacial acetic acid after 16 hours at 0°C. The polymer was precipitated in a tenfold excess of methanol and dried in a vacuum oven overnight at 60°C.

Characterization

1. *Gel permeation chromatography*. Gel permeation chromatography (GPC) was performed on a Waters Model 200 apparatus using methylene chloride as the solvent at 25°C. Polymer solutions, 0.5 percent wt/vol, were injected for 50 seconds into a 1 ml/min flow stream. Styragel®-packed columns of 10^6, 10^5, 10^4, 10^3, and 10^2 "angstrom" nominal porosities were employed.

2. *Number average molecular weight*. Number average molecular weights $(\overline{M}_n)$ were determined on a Hallikainen 1361-B membrane osmometer in benzene at 37°C.

3. *Intrinsic viscosity*. Intrinsic viscosities of the PMMAs were determined in Ubbelohde-type capillary viscometers in $CHCl_3$ at 25°C.

4. *Nuclear magnetic resonance spectroscopy*. NMR spectra of the poly(methyl

methacrylates) were recorded on a modified Varian HA-100 nuclear magnetic resonance spectrometer. Deuterated chloroform containing 1 percent tetramethylsilane was used as the solvent. The spectra were analyzed according to the procedure of Bovey and Tiers [74] for the microtacticity of the PMMA.

Polymers of CMPT were examined in THF solution, with cyclohexane as an internal standard. The following assignments have been made for the silicon methyl group triads: syndio, 9.91 τ; hetero, 9.93 τ; and iso, 9.96 τ [25,75].

5. *Electron spin resonance spectroscopy.* ESR spectra of samples in capillary tubes were recorded on a Jelco JES-ME-1X electron spin resonance spectrometer at room temperature and at a frequency of 9,400 MHz using a magnetic field modulation of 100 KHz. The overmodulated spectra were doubly integrated by the Wyard method [76] to calculate the concentration of free radicals in solution.

6. *Electron microscopy.* Electron microscopy was performed on Philips EM-100B electron microscope. A 1 percent solution of the block polymer in methylene chloride was cast on a clean glass slide. The dry, thin film (~500–700 Å) was floated off the glass slide onto distilled water, supported on an EM grid, stabilized with a thin layer of carbon, and used for direct transmission micrographs.

7. *Foam fractionation.* Block polymer 98-1 (HI) was foam-fractionated [77] using the apparatus described by Gaines and Bender [48]. A solution of 4.034 g of this block polymer in one liter of methylene chloride was foamed, and six ~100-ml foam fractions were collected. The foam fractions were isolated by solvent evaporation and drying in a vacuum oven. These fractions were then analyzed for percent dimethylsiloxane, and the IVs were measured.

8. *Silicon analysis.* A sample was fused with Na_2O_2 in a Parr bomb, and the fusion product was dissolved in 100 ml of a 50 percent aqueous $HClO_4$ solution. The solution was dehydrated until the $HClO_4$ fumed. The precipitated SiO_2 was weighed, and the weight-percent dimethylsiloxane was calculated with the appropriate conversion factor (1.23 × weight-percent SiO_2). Duplicate analyses for weight-percent D agreed to within ±0.4 percent.

Results and Discussion

Overall Approach

The ambident-initiator approach for the synthesis of acrylic–siloxane block polymers is shown in Figure 1. The dilithium benzophenone dianion possesses two sites of unequal reactivity toward D_3: one a sterically hindered 1,1-diphenylmethyl carbanion, the other a relatively unhindered alkoxy anion. The effect of

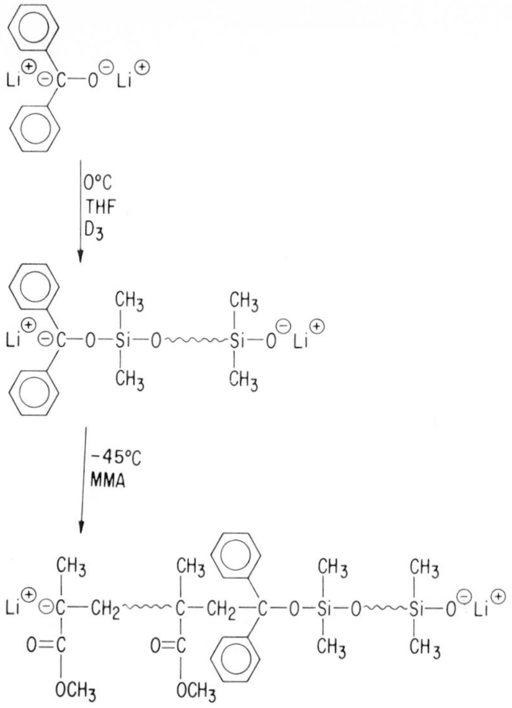

Figure 1. Ambident-initiator approach for the synthesis of acrylic–siloxane block polymers.

steric hindrance is amplified by the fact that a THF–toluene solution of polystyryl lithium and D_3 must be heated to 50°C to affect the crossover reaction [36].

Another property of this class of initiators is their propensity to transfer an electron to substrates with high electron affinities [78]. Thus, for example, when MMA or HMPA are added to BPDA solutions, the dianion transfers an electron, as shown in Equation (1), and is transformed to the ketyl

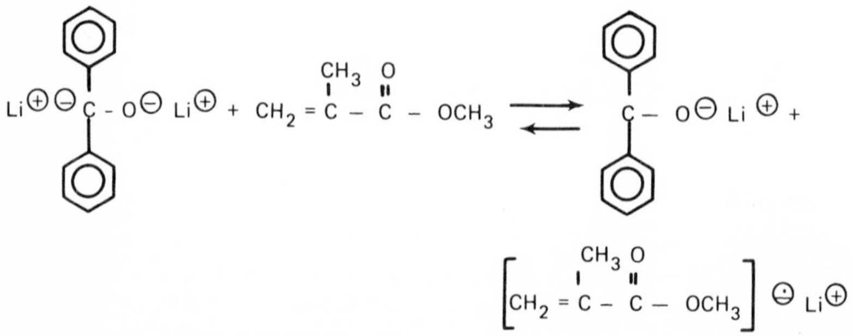

$$ (1) $$

However, D_3, with a relatively low electron affinity, adds preferentially to the alkoxy anion. This fact is supported by the ESR spectra of BPDA before and

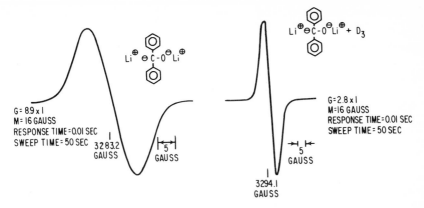

Figure 2. ESR spectra of BPDA and BPDA containing D_3.

after the addition of D_3 (Figure 2). The ESR signal measured is that of lithium benzophenone ketyl, which was present at a level of <0.5 percent in the BPDA. When these spectra were doubly integrated [76], normalized for the different amplification factors (G = gain, M = modulation width, [Figure 2]), and corrected for dilution by D_3, nearly identical lithium benzophenone ketyl concentrations were calculated.

Addition of D_3 to BPDA in THF also resulted in a color change of the solution from the red-purple of the dianion to the orange-brown of the 1,1-diphenyl-methyl-carbanion-terminated siloxane chain. Shown in Figure 3 are the absorption maxima of several 1,1-diphenylmethyl-derived anions in THF at 25°C.

The final step shown in Figure 1 is the polymerization of MMA by the terminal 1,1-diphenylmethyl carbanion of the polydimethylsiloxane chain. This step

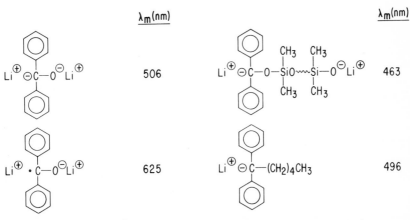

Figure 3. Absorption maxima of several 1,1-diphenylmethyl-derived anions.

can be conducted in the presence of the lithium siloxanolate chain-ends because these species cannot initiate the polymerization of MMA.

Initiation Reaction and D_3 Polymerization

The initiation reaction of D_3 with dilithium benzophenone dianion requires one polymer chain to be produced per initiator molecule (Figure 1). Thus, the predicted number average molecular weight, $\overline{M}_k$, is given by:

$$\overline{M}_k = w/I, \tag{2}$$

where w is the weight of D_3 and I is the number of moles of BPDA. Figure 4 shows a test of the stoichiometry predicted by Equation (2). In all cases, the $\overline{M}_n$ measured by GPC was higher than the predicted $\overline{M}_k$. Furthermore, the molecular-weight distributions (MWD) were $\geqslant 1.35$. These two results can be explained in terms of a slow reaction of the 1,1-diphenylmethyl carbanion with D_3. Further evidence that this reaction had taken place is presented below.

The presence of initiator residues in the polysiloxanes produced with BPDA is

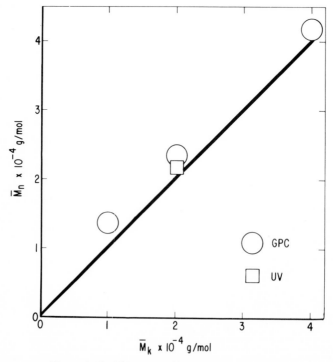

Figure 4. Stoichiometry of the initiation reaction.

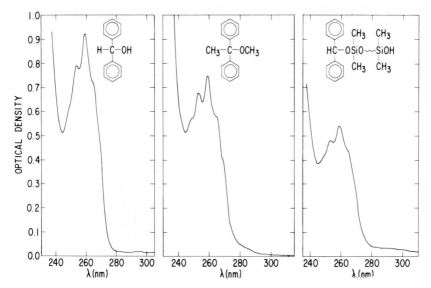

Figure 5. Initiator residues in polydimethylsiloxanes prepared with BPDA.

shown in Figure 5. Polydimethylsiloxanes are transparent in this region of the UV spectrum.

When the D_3 polymerization with BPDA was allowed to continue for ~20 hours at 0°C, the onset of gelation was observed. The orange-brown gel became completely immobile after ~24 hours at 0°C. This gel could be broken by the addition of equimolar amounts (1:1 based on Li) of a monofunctional salt (ϕ_3SiOLi · HMPA), a stronger solvating agent than THF (HMPA), or a proton donor (glacial acetic acid). These observations support the conclusion that gel formation was due to the cross-association [22] of the lithiated chain-ends (Figure 1).

Block Polymer Synthesis and Characterization

In the initial experiments, D_3 was polymerized with BPDA for 24 hours at 0°C, after which time the gel was broken with HMPA (1:1 based on Li). MMA was then added at -45°C; the block polymers obtained are listed in Table I. The GPCs of the hexane-insoluble portions of these block polymers are shown in Figure 6.

These block polymers were synthesized under identical polymerization conditions and demonstrated the lack of reproducibility in this system. Moreover, a methylene-chloride-cast film of block polymer 98-1 (HI) was optically clear, whereas block polymers 102E (HI) and 123-1 (HI) produced opaque films when cast under identical conditions.

TABLE I

Effect of PMMA Molecular-Weight Distribution

Sample	Yield (%)	$[\eta]$ (dl/g)	Hexane Insoluble		
			%	$[\eta]$ (dl/g)	% D
98-1	98.4	0.47	86.5	0.49	24.9
102E	97.2	0.89	85.2	0.88	22.2
123-1	94.9	0.72	78.8	0.89	15.3

Note: $[BPDA]_o = 3.04 \times 10^{-3}$ M, $[D_3]_o = 2.74 \times 10^{-4}$ M, $[MMA]_o = 1.52 \times 10^{-3}$ M, THF = 150 ml.

The lack of reproducibility with this system was probably due to the inability to determine when all the D_3 was consumed, even though an apparent "gel" point was reached. Any remaining D_3 could react with a growing PMMA chain and terminate it prematurely. Since high yields of block polymers were obtained in all cases, the monomer apportioned for the prematurely terminated chain could add to other propagating PMMA chains, thus broadening the PMMA molecular-weight distribution. The broadened PMMA MWD is readily shown in the GPC traces by the positive response of the detector at high molecular weights. Since polydimethylsiloxanes produce a negative detector response in CH_2Cl_2, the ac-

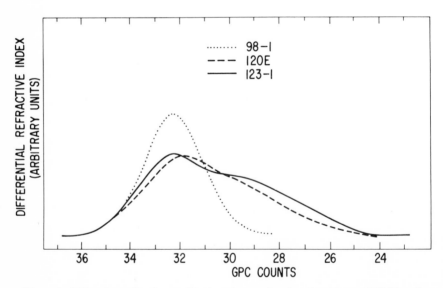

Figure 6. GPCs of acrylic–siloxane block polymers showing the effect of PMMA molecular-weight distribution.

tual height of the trace would be reduced in proportion to the siloxane content of the high-molecular-weight block polymer.

The broadened PMMA MWD also affects the *composition distribution* of the block polymers. That a composition distribution existed even in block polymer 98-1 (HI), which produced a clear film, is shown by the results of a foam fractionation [48,74] (Table II). The composition of the fractions of this block polymer ranged from 39.4 to 2.1 weight-percent of combined dimethylsiloxy (D) units. By contrast, the first foam fraction of block polymer 102E (HI) contained more than 60 percent D and was a greasy solid.

Thus, a controlled level of compositional heterogeneity is not detrimental to the optical properties of solvent-cast block polymer films. Broader composition distributions, as obtained with block polymers 102E (HI) and 123-1 (HI), probably invite gross phase separation and produce opaque solvent-cast films.

Another feature of this system is shown by the data in Table I, namely, the presence of a hexane-soluble portion of the block polymers. The IR spectra showed that little PMMA was bound to the essentially pure polydimethylsiloxane that was extracted. This observation leads to the conclusion that the 1,1-diphenylmethyl carbanion is not completely hindered in its reaction with D_3. From the weight fractions of hexane-soluble and hexane-insoluble polymer and their compositions, we estimate that 20 to 40 percent of the 1,1-diphenylmethyl carbanion disappeared by reaction with D_3 prior to the addition of MMA. This result also explains the broadening of the polydimethylsiloxane MWD and the lack of agreement of $\overline{M}_n$ with $\overline{M}_k$ described above.

In an attempt to provide a reproducible synthesis of acrylic–siloxane block polymers, a "seeding" technique was employed (Figure 7). In this modified synthesis, the D_3 was initiated completely with BPDA (30–45 minutes at $0°C$), and then HMPA was added. The presence of HMPA increased the rate of the D_3 polymerization (8 hours versus 24 hours) and prevented gel formation at complete conversion. The block polymers obtained with the modified synthesis are shown in Table III.

TABLE II

Composition Distribution of Block Polymer 98-1 (HI)
by Foam Fractionation

Fraction	Wt-%	% D	[η]
1	14.8	39.4	0.51
2	15.8	38.1	0.41
3	20.2	32.8	0.40
4	15.1	22.0	0.51
5	8.1	12.4	0.45
6	26.1	2.1	0.50

Note: $\Sigma w_i D_i = 23.3$; $\Sigma \eta_i w_i = 0.47$.

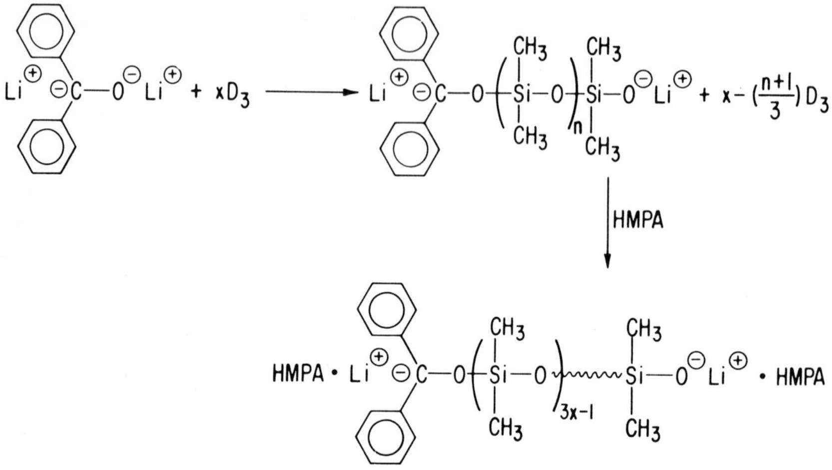

Figure 7. Modified synthesis of acrylic–siloxane block polymers.

These results demonstrate that a reproducible block polymer synthesis was achieved. Reduction of the initiator level by a factor of two produced block polymers with two times the molecular weight and nearly identical polydimethylsiloxane contents. The GPCs of these block polymers (hexane-insoluble portions) (Figure 8) did not reveal any high-molecular-weight tails. Moreover, all of these block polymers produced optically clear films when cast from methylene chloride.

The seeded polymerization technique was extended to the synthesis of a series of block polymers with a constant polydimethylsiloxane block length, $\bar{M}_k \approx$ 20,000 g/mol (Table IV). To provide different polydimethylsiloxane-containing block polymers, the PMMA block molecular weight was reduced from design values of 113,000 g/mol for 69-1 to 800 g/mol for 39-1.

TABLE III

Reproducibility of the Modified Synthesis of Acrylic–Siloxane Block Polymers

Sample	Yield	% D	Hexane Insolubles			
			$\bar{M}_n \times 10^{-4}$ (g/mol)	%	$[\eta]$ (dl/g)	% D
34-1	99.3	35.2	8.27	87.7	0.66	21.6
11-9	94.0	32.0	–	86.8	0.73	22.7
54-4	97.7	34.3	–	85.4	0.66	21.7
18-1	97.5	34.6	4.17	84.5	0.44	21.7
25-1	95.9	38.9	–	86.4	0.44	23.5
31-1	97.9	34.6	–	82.7	0.46	21.4

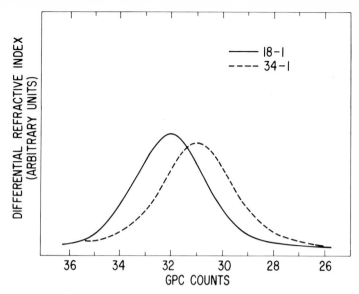

Figure 8. GPCs of block polymers 18-1 (HI) and 34-1 (HI).

The amount of hexane-extractable material increased as the polydimethyl-siloxane content increased. Unlike the results obtained with block polymer 98-1, the hexane-soluble portions of block polymers 83-1, 38-1, 38-2, and 38-3 contained PMMA (IR spectroscopy). The presence of PMMA in these hexane-soluble fractions was probably due to the shorter PMMA blocks, which, by virtue of their attachment to siloxane chains, were rendered soluble.

A final consideration in this synthetic method is the effect of HMPA on the

TABLE IV

Acrylic–Siloxane Block Polymers with a Constant Polydimethylsiloxane Block Length

Sample	% D		Yield	Hexane Insoluble	
	Design	FD		%	% D
69-1	15	15.6	88.0	98.1	14.3
22-1	35	34.3	97.5	86.8	24.8
83-1	50	51.3	95.5	76.6	34.1
38-1	55	58.1	94.3	75.0	44.6
38-2	62	60.2	85.4	73.7	48.5
38-3	71	74.7	86.3	62.0	58.1
38-4	83	85.0	83.0	swollen	
39-1	96	96.7	87.0	soluble	

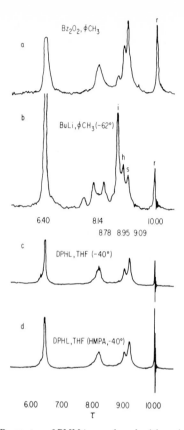

Figure 9. NMR spectra of PMMAs produced with various initiators.

microtacticity of the PMMA and on the course of the D_3 polymerization, that is, if equilibration had occurred with the latter.

Shown in Figure 9 are NMR spectra of PMMAs produced with various initiators [74]. Also included in the figure are the NMR spectra of PMMA produced with DPHL in THF with and without HMPA (Figures 9c and 9d, respectively). A comparison of these latter spectra reveals that HMPA (1:1 based on Li) does not affect the predominantly syndiotactic triad content of PMMA. Thus, the polarity of the medium with THF solvent is sufficient to influence the MMA monomer placement [79]. The influence of any specific solvation effect on the microtacticity of PMMA with HMPA and Li counterions in THF is ruled out.

An indirect method [25] was employed to examine whether or not HMPA promoted equilibration (broadened MWD, higher cyclics) during the D_3 polymerization with BPDA. In this method cis-2,4,6-trimethyl-2,4,6-triphenylcyclotrisiloxane (CMPT) was polymerized with BPDA, and BPDA solvated with HMPA (1:1 based on Li) in THF. The silicon methyl regions of the NMR spec-

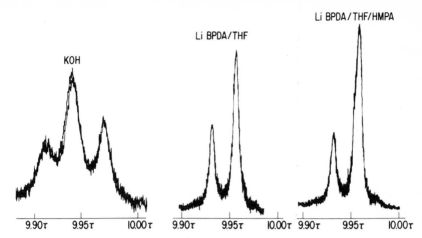

Figure 10. Silicon methyl proton regions of NMR spectra of CMPT polymers.

tra of the polymers are shown in Figure 10. Included for comparison is the same region of a CMPT polymer prepared with KOH.

The stereospecific ring opening (S_Ni–Si) [80] of CMPT produced a poly(methylphenylsiloxane) containing ~67 percent isotactic and ~33 percent heterotactic silicon methyl triads [25]. These triad distributions were observed with Li-based initiators that did not promote equilibration or siloxane–siloxanolate redistributions [25]. With equilibration catalysts such as KOH, the following triad contents were measured: syndiotactic ~25 percent, heterotactic ~50 percent, and isotactic ~25 percent [25]. The results shown in Figure 10 demonstrate conclusively that HMPA-solvated BDPA does not promote equilibration in the CMPT system and, by extension, in the D_3 system.

Block Polymer Properties

1. *Clarity*. The light transmission of two acrylic–siloxane block polymer films (10 mil, cast from methylene chloride) is compared with that of a similarly prepared film of a commercial PMMA, Lucite® 40, in Table V. These block

TABLE V

Light Transmission of Block
Polymer Films

Polymer	% D	% T
Lucite® 40	0	94.0
AS-54-4	21.7	93.5
AS-83-1	34.1	94.0

polymers contain 21.7 [54-4 (HI)] and 34.1 [83-1 (HI)] percent dimethylsiloxane by weight. The clarity of these block polymers cannot be due to a refractive
index match of the two components (n = 1.491 PMMA [81]; n = 1.404 PDMS
[82]) or to a mutual solubility of the two components [δ (cal/cc)$^{1/2}$ = 9–9.5,
PMMA; δ = 7.3–7.6, PDMS] [83]. Another possibility is that the polydimethylsiloxane regions are microdispersed in a PMMA matrix and that the dimensions
of the dispersed regions are less than 500 Å [84,85]. The results of transmission
electron microscopy are shown in Figures 11–13 and support the conclusion that
the clarity of the block polymers is due to microdispersed polydimethylsiloxane
regions. These transmission photomicrographs were obtained on unstained samples, the contrast being provided by the electron density differences between the
two components [34,47]. In Figure 13, the polydimethylsiloxane regions are
shown to be deformed in the vicinity of the hole in the film. Deformed polydimethylsiloxane regions are evident for ~2,000 Å away from the edge of the
hole.

2. *Surface activity.* Mixtures of block polymer 34-1 (HI) and PMMA
(Lucite® 40) were cast from methylene chloride into ~10 mil films. The water

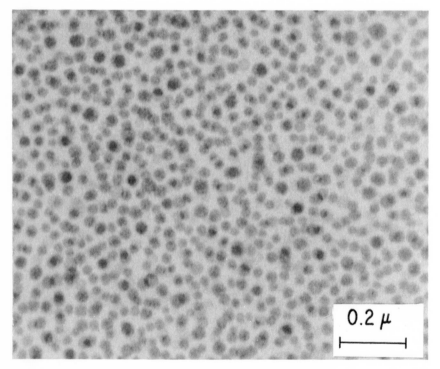

Figure 11. Electron micrograph of block polymer 54-4 (HI) containing 21.7 wt-% D.
× 85K.

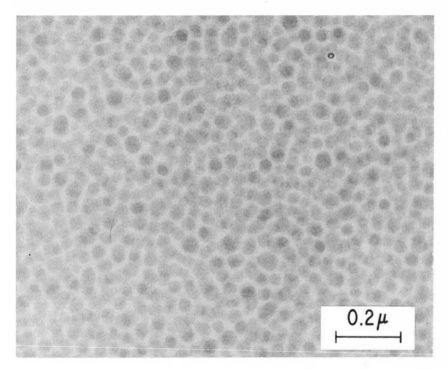

Figure 12. Electron micrograph of block polymer 83-1 (HI) containing 34.1 wt-% D.
× 85K.

TABLE VI

Water Contact Angles of PMMA Films Containing
Acrylic–Siloxane Block Polymer

Block Polymer in PMMA (wt-%)	Water Contact Angle (°)
0	61
10^{-3}	75
10^{-2}	87
10^{-1}	95
1	97
10	95
50	95
99	95

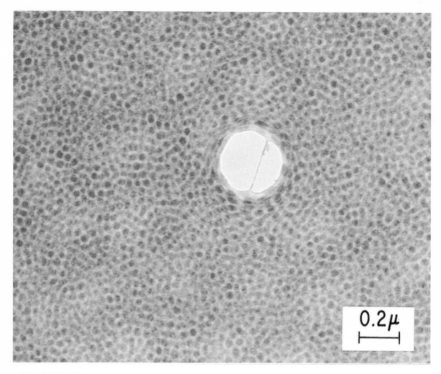

Figure 13. Electron micrograph of block polymer 54-4 (HI) showing deformation of poly-dimethylsiloxane regions. X 50K.

contact angles of these films were measured on the air–polymer interface surface with a low-power microscope having a protractor eyepiece. These results are presented in Table VI. A very low concentration ($\sim 10^{-1}$ weight-percent) of block polymer in PMMA is effective in producing a polydimethylsiloxane surface. This result parallels those found with other organic polymer/dimethylsiloxane–organic block polymer mixtures [46,48,86].

Summary

The design features for the synthesis of well-defined organic–siloxane block polymers with organometallic initiators have been described. A detailed description of the synthesis and properties of acrylic–siloxane block polymers has been presented.

Acknowledgment

The authors acknowledge the many helpful discussions with Drs. E. E. Bostick, J. W. Dean, W. A. Fessler, G. L. Gaines, Jr., S. W. Kantor, and D. G. LeGrand during the course of this study. Excellent analytical support was provided by the Materials Characterization Operation, General Electric Corporate Research and Development.

References

1. U.S. Patent No. 2,490,357, December 6, 1949, "Polymerization of Siloxanes," James F. Hyde to the Corning Glass Works, Corning, New York.
2. U.S. Patent No. 2,567,110, September 4, 1951, "Organopolysiloxanes Prepared by the Reaction of Salts of Silanols with Halosilanes," James F. Hyde to the Dow Corning Corporation, Midland, Michigan.
3. Grubb, W.T. and Osthoff, R.C., "Kinetics of the Polymerization of a Cyclic Dimethylsiloxane," *J. Amer. Chem. Soc.*, 77, no. 6 (1955), 1405.
4. Kucera, M., "Base-catalyzed Polymerization of Octamethylcyclotetrasiloxane. II. Kinetics of the Reaction," *Coll. Czech. Chem. Commun.*, 25, no. 2 (1960), 547.
5. Kucera, M., "Changes in the Rate of Anionic Polymerization of Octamethylcyclotetrasiloxane Caused by the Formation of Complexes on the Silanolate Active Centers," *J. Polym. Sci.*, 58, no. 166 (1962), 1263.
6. Laita, Z. and Jelinek, M., "The Kinetics of the Anionic Polymerization of Cyclic Polydimethylsiloxanes," *Polym. Sci. USSR*, 4, no. 3 (1963), 535.
7. Scott, D.W., "Equilibria Between Linear and Cyclic Polymers in Methylpolysiloxanes," *J. Amer. Chem. Soc.*, 68, no. 11 (1946), 2294.
8. Hurd, D.T., Osthoff, R.C., and Corrin, M.L., "The Mechanism of Base-catalyzed Rearrangement of Organopolysiloxanes," *J. Amer. Chem. Soc.*, 76, no. 1 (1954), 249.
9. Kantor, S.W., Grubb, W.T., and Osthoff, R.C., "The Mechanism of the Acid- and Base-catalyzed Equilibration of Siloxanes," *J. Amer. Chem. Soc.*, 76, no. 20 (1954), 5190.
10. Kucera, M. and Jelinek, M., "Base-catalyzed Polymerization of Octamethylcyclotetrasiloxane. I. Mechanism of the Reaction," *Coll. Czech. Chem. Commun.*, 25, no. 2 (1960), 536.
11. Kucera, M. and Jelinek, M., "Chain Transfer in the Anionic Polymerization of Octamethylcyclotetrasiloxane," *Polym. Sci. USSR*, 3, no. 3 (1962), 514.
12. U.S. Patent No. 2,634,284, April 7, 1953, "Polymerization of Organosiloxanes," James F. Hyde to the Dow Corning Corporation, Midland, Michigan.
13. Cooper, G.D. and Elliot, J.R., "Promotion of Base-catalyzed Siloxane Rearrangements by Dimethyl Sulfoxide," *J. Polym. Sci., Part A-1*, 4, no. 3,(1966), 603.
14. Ostrozynski, R.L., "Phosphine Oxide Promoted Anionic Induced Siloxane Rearrangements," *Polym. Prep., Amer. Chem. Soc., Div. Polym. Chem.*, 8, no. 1 (1967), 474.
15. Yuzhelevskii, Yu. A., Kagan, Ye. G., and Fedoseyeva, N.N., "Activators of the Anionic Polymerization of Cyclosiloxanes," *Polym. Sci. USSR*, 12, no. 7 (1970), 1800.
16. Bostick, E.E., "VII. Interchange Reactions B. Silicones," in *Chemical Reactions of*

Polymers, Vol. 19, High Polymers, E.M. Fettes, ed., New York: Interscience (1964), 515.

17. Noll, W., Chemistry and Technology of Silicones, New York: Academic Press (1968), 226.

18. U.S. Patent No. 3,051,684, August 28, 1962, "Organosiloxane Polymers Containing Polydiene Blocks," Maurice Morton and Alan Rembaum to the Board of Directors of the University of Akron, Ohio.

19. Morton, M., Rembaum, A.A., and Bostick, E.E., "Block Copolymerization of Unsaturated Monomers and Octamethylcyclotetrasiloxane," J. Appl. Polym. Sci., 8, no. 11 (1964), 2707.

20. Mitoh, M., Tabuse, A., and Minoura, U., "Anionic Copolymerization of Octamethylcyclotetrasiloxane and Styrene," Kogyo Kagaku Zasshi, 70, no. 11 (1967), 1969.

21. Minoura, Y., Mitoh, M., Tabuse, A., Yamada, Y., "Copolymerization of Octamethylcyclotetrasiloxane with Vinyl Compounds. II. Synthesis of Block Copolymers of Dimethylsiloxane and Vinyl Compounds," J. Polym. Sci., Part A-1, 7, no. 9 (1969), 2753.

22. Szwarc, M., Carbanions, Living Polymers, and Electron-transfer Processes, New York: Interscience (1968), 27–103, 22a, 476–519.

23. U.S. Patent No. 3,337,497, August 22, 1967, "Polysiloxane Block Copolymers," Edgar E. Bostick to General Electric Company, Schenectady, New York.

24. Lee, C.L., Frye, C.L., and Johannson, O.K., "Selective Polymerization of Reactive Cyclosiloxanes to Give Non-equilibrium Molecular Weight Distributions. Monodisperse Siloxane Polymers," Polym. Prep., Amer. Chem. Soc., Div. Polym. Chem., 10, no. 2 (1969), 1361.

25. Bostick, E.E. and Bush, J.B., Jr., "Stereochemistry of Polymerization of Methylphenylcyclosiloxanes," Résumés des communications, 11e symposium international sur la chimie des composés organiques du silicium, Bordeaux, July 9–12, 1968, 24.

26. U.S. Patent No. 3,481,898, December 2, 1969, "Polymerization of Siloxanes," William G. Davies, Brian Elliot, and Thomas C. Kendrick to Midland Silicones Ltd., Reading, England.

27. Frye, C.L., Salinger, R.M., Fearon, F.W.G., Klosowski, J.M., and DeYoung, T., "Reactions of Organolithium Reagents with Siloxane Substrates," J. Org. Chem., 35, no. 5 (1970), 1308.

28. Juliano, P.C., Fessler, W.A., and Cargioli, J.D., "Proton Magnetic Resonance Study of Lithium Silanolates. I. Preliminary Observations," Macromol. Prepr., XIII Cong., Int. Union Pure Appl. Chem., 2 (1971), 1212.

29. Juliano, P.C., Fessler, W.A., and Cargioli, J.D., "Proton Magnetic Resonance Study of Lithium Silanolates. II. Effects of Solvating Agent Stoichiometry and Temperature," Polym. Prepr., Amer. Chem. Soc., Div. Polym. Chem., 12, no. 2 (1971), 158.

30. Fessler, W.A. and Juliano, P.C., "Reactivity of Solvated Lithium n-Butyldimethylsilanolate with Organosiloxane Substrates," Polym. Prepr., Amer. Chem. Soc., Div. Polym. Chem., 12, no. 2 (1971), 150.

31. Bostick, E.E., "Synthesis and Properties of Siloxane Block Polymers," in Block Polymers, S.W. Aggarwal, ed., New York: Plenum Press (1970), 237.

32. German Offen. 2,049,547, April 22, 1971, "Organopolysiloxane Block Polymers," Edgar E. Bostick and William A. Fessler to the General Electric Company, Schenectady, New York.

33. U.S. Patent No. 3,483,270, December 9, 1969, "Stereospecific Organosiloxane–Macromolecular Hydrocarbon Block Copolymers," Edgar E. Bostick to the General Electric Company, Schenectady, New York.

34. Saam, J.C., Gordon, D.J., and Lindsey, S., "Block Copolymers of Polydimethylsiloxane and Polystyrene," *Macromolecules,* 3, no. 1 (1970), 1.

35. German Offen. 2,011,088, September 24, 1970, "Block Copolymers from Aromatic Organic Blocks and Diorganosiloxane Blocks," J.C. Saam to the Dow Corning Corporation, Midland, Michigan.

36. Dean, J.W., "Multiple-Sequence Block Polymers of Poly(styrene) and Poly(dimethylsiloxane)," *J. Polym. Sci., Part B,* 8, no. 10 (1970), 677.

37. Davies, W.G. and Jones, D.P., "Styrene–Siloxane ABA Block Copolymers," *Ind. Eng. Chem., Prod. Res. Develop.,* 10, no. 2 (1971), 168.

38. French Demande 2,046,978, April 23, 1971, "Sequenced Organosiloxane Copolymers," W.G. Davies, T.C. Kendrick, and D.P. Jones to Midland Silicones Ltd., Reading, England.

39. German Offen. 2,116,837, October 21, 1971, "Hexamethylcyclotrisiloxane-styrene Block Copolymer," John W. Dean to the General Electric Company, Schenectady, New York.

40. German Offen. 2,142,594, March 2, 1972, "Thermoplastic Poly(diorganosiloxane)–Polystyrene Block Copolymers," J.C. Saam and F.W.G. Fearon to the Dow Corning Corporation, Midland, Michigan.

41. German Offen. 2,142,595, March 2, 1972, "Thermoplastic Siloxane–styrene Block Copolymers," J.C. Saam and F.W.G. Fearon to the Dow Corning Corporation, Midland, Michigan.

42. U.S. Patent No. 3,678,125, July 18, 1972, "Siloxane Containing Thermoplastic," J.C. Saam and F.W.G. Fearon to the Dow Corning Corporation, Midland, Michigan.

43. U.S. Patent No. 3,678,126, July 18, 1972, "Siloxane Containing Thermoplastic Elastomers," J.C. Saam and F.W.G. Fearon to the Dow Corning Corporation, Midland, Michigan.

44. German Offen. 2,142,664, March 2, 1972, "Copolymers with Poly(alpha-methylstyrene) and Poly(diorganosiloxane) Blocks," J.C. Saam and F.W.G. Fearon to the Dow Corning Corporation, Midland, Michigan.

45. U.S. Patent No. 3,665,052, May 23, 1972, "Method of Preparing a Block Copolymer Containing a Poly-alpha-methylstyrene Block and a Polydiorganosiloxane Block," J.C. Saam and F.W.G. Fearon to the Dow Corning Corporation, Midland, Michigan.

46. Owen, M.J. and Kendrick, T.C., "Surface Activity of Polystyrene–Polysiloxane–Polystyrene ABA Block Copolymers," *Macromolecules,* 3, no. 4 (1970), 458.

47. Saam, J.C. and Fearon, F.W.G., "Properties of Polystyrene–Polydimethylsiloxane Block Copolymers," *Ind. Eng. Chem., Prod. Res. Develop.,* 10, no. 1 (1971), 10.

48. Gaines, G.L., Jr., and Bender, G.W., "Surface Concentration of Styrene–Dimethylsiloxane Block Copolymer in Mixtures with Polystyrene," *Macromolecules,* 5, no. 1 (1972), 82.

49. British Patent No. 1,257,304, March 29, 1968, "Polymer Compositions," Thomas C. Kendrick and Michael J. Owen to Midland Silicons Ltd., Reading, England.

50. British Patent No. 1,261,301, February 11, 1970, "Blends of Organosiloxane Gums and Block Copolymers of Polyvinyl Aromatics and Polydimethylsiloxanes," J.C. Saam and Charles W. Lentz to the Dow Corning Corporation, Midland, Michigan.

51. British Patent No. 1,261,484, February 12, 1970, "Polymer Blend," J.C. Saam to the Dow Corning Corporation, Midland, Michigan.

52. German Offen. 2,011,046, September 17, 1970, "Curable Organopolysiloxane Elastomer Compositions," J.C. Saam and C.W. Lentz to the Dow Corning Corporation, Midland, Michigan.

53. German Offen. 2,011,059, September 17, 1970, "Molded Articles and Coatings of Ther-

mosetting Mixtures Based on Vinyl Aromatic Polymers," J.C. Saam to the Dow Corning Corporation, Midland, Michigan.

54. German Offen. 2,116,836, October 21, 1971, "Polystyrene Composition Containing Dimethylsiloxane–styrene Block Copolymers," John W. Dean to the General Electric Corporation, Schenectady, New York.

55. German Offen. 2,142,597, March 7, 1972, "Homogeneously Mixing Organic Vinyl Thermoplastics and Polyorganosiloxanes," J.C. Saam to the Dow Corning Corporation, Midland, Michigan.

56. U.S. Patent No. 3,663,650, May 16, 1972, "Substantially Transparent Polydimethylsiloxane–Polyalkylmethacrylate Compositions and Methods for Making Same," Peter C. Juliano to the General Electric Company, Schenectady, New York.

57. Braun, D. and Neumann, W., "Polymerisationsauslosung durch Radikalanionen und Dianionen aus ungesättigten Verbindungen," *Macromol. Chem.*, 92,(1966), 180.

58. Panaiotov, I., Tsvetanov, C., and Ivanov, S., "Vinyl Polymerization Initiated by the Benzil Anion Radical and the Benzil Dianion," *C.R. Acad. Bulg. Sci.*, 20, no. 9 (1967), 927. (In Russian)

59. U.S. Patent No. 3,410,836, November 12, 1968, "Polymerization of Conjugated Dienes with a Dilithium Complex of an Aromatic Ketone," Henry L. Hsieh and William J. Trepka to Phillips Petroleum Company, Bartlesville, Oklahoma.

60. Sumitomo, H. and Hashimoto, K., "Polymerization of β-Cyanopropionaldehyde. V. Anionic Polymerization Initiated by Benzophenone–Alkali Metal Complexes," *J. Polym. Sci., Part A-1*, 7, no. 5 (1969), 1331.

61. Minoura, Y. and Tsuboi, S., "Polymerization of Vinyl Monomers by Alkali Metal–Thiobenzophenone Complexes," *J. Polym. Sci., Part A-1*, 8, no. 1 (1970), 125.

62. Hashimoto, K. and Sumitomo, H., "Polymerization of β-Cyanopropionaldehyde. X. Anionic Copolymerization with Methyl Isocyanate," *J. Polym. Sci., Part A-1*, 9, no. 1 (1971), 107.

63. Glusker, D.L., Stiles, E., and Yankoski, B., "The Mechanism of Anionic Polymerization of Methyl Methacrylate. I. Quantitative Determination of Active Chains Using Radioactive Terminators," *J. Polym. Sci.*, 49, no. 152 (1961), 292.

64. Glusker, D.L., Lysloff, I., and Stiles, E., "The Mechanism of Anionic Polymerization of Methyl Methacrylate. II. The Use of Molecular Weight Distributions to Establish a Mechanism," *J. Polym. Sci.*, 49, no. 152 (1961), 315.

65. Wiles, D.M. and Bywater, S., "The Butyllithium-initiated Polymerization of Methyl Methacrylate," *Polymer*, 3, no. 1 (1962), 175.

66. Cottam, B.J., Wiles, D.M., and Bywater, S., "The Butyllithium-initiated Polymerization of Methyl Methacrylate. Part II. Molecular Weight Distributions," *Can. J. Chem.*, 41, no. 8 (1963), 1905.

67. Wiles, D.M., "Polymerization of α,β-Unsaturated Carbonyl Compounds," in *Structure and Mechanism in Vinyl Polymerization*, T. Tsuruta and K.F. O'Driscoll, eds., New York: Marcel Dekker (1969), 223.

68. Wiles, D.M. and Bywater, S., "Methoxide Ions in the Anionic Polymerization of Methyl Methacrylate," *Chem. Ind. (London)*, no. 29 (1963), 1209.

69. Patnode, W.I. and Wilcock, D., "Methylpolysiloxanes," *J. Amer. Chem. Soc.*, 68, no. 3 (1946), 358.

70. Young, C.W., Servais, P.C., Currie, C.C., and Hunter, M.J., "Organosilicon Polymers. IV. Infrared Studies on Cyclic Substituted Siloxanes," *J. Amer. Chem. Soc.*, 70, no. 11 (1948), 3758.

71. Hickton, H.J., Holt, A., Homer, J., and Jarvie, A.W., "Organosilicon Compounds, Part 1. The 2,4,6-Trimethyl-2,4,6-triphenyl-cyclotrisiloxanes and the 2,4,6,8-Tetramethyl-2,4,6,8-tetraphenylcyclotetrasiloxanes," *J. Chem. Soc.*, C, no. 2 (1966), 149.

72. Evans, A.G. and George, D.B., "The Catalytic Action of Anionic Catalysts, Part 1. The Interaction of Butyllithium with 1,1-Diphenylethylene," *J. Chem. Soc.,* no. 910 (1961), 4653.

73. Waack, R., Doran, M.A., and Stevenson, P.E., "Solvent and Concentration Effects on the Rate of Addition of *n*-Butyllithium to 1,1-Diphenylethylene," *J. Organometal. Chem.,* 3 (1965), 481.

74. Bovey, F.A. and Tiers, G.V.D., "Polymer NSR Spectroscopy. II. The High Resolution Spectra of Methyl Methacrylate Polymers Prepared with Free Radical and Anionic Initiators," *J. Polym. Sci.,* 44, no. 143 (1960), 173.

75. Fessler, W.A., private communication.

76. Wyard, S.J., "Double Integration of Electron Spin Resonance Spectra," *J. Sci. Instrum.,* 42, no. 10 (1965), 769.

77. Gaines, G.L., Jr., and LeGrand, D.G., "Foam Fractionation of Polymers," *J. Polym. Sci., Part B,* 6, no. 9 (1968), 625.

78. McClelland, B.J., "Anionic Free Radicals," *Chem. Rev.,* 64, no. 3 (1964), 301.

79. Roig, A., Figueruelo, J.E., and Llano, E., "Monodisperse Stereoregular Poly(methyl methacrylate) by Anionic Polymerization. II," *J. Polym. Sci., Part C,* no. 16 (1968), 4141.

80. Sommer, L.H., *Stereochemistry, Mechanism and Silicon,* New York: McGraw-Hill (1965), 48.

81. E. I. DuPont de Nemours, *Lucite Acrylic Resin-Design Handbook* (1968), 13.

82. McGregor, R.R., *Silicones and Their Uses,* New York: McGraw-Hill (1954), 38.

83. Brandup, J. and Immergut, E.H., *Polymer Handbook,* New York: Interscience (1966), iv–364.

84. Estes, G.M., Cooper, S.L., and Tobolsky, A.V., "Block Polymers and Related Heterophase Elastomers," *J. Macromol. Sci., Pt. C, Rev. Macromol. Chem.,* 4, no. 2 (1970), 313.

85. Morton, M., McGrath, J.E., and Juliano, P.C., "Structure–Property Relationships for Styrene–Diene Thermoplastic Elastomers," *J. Polym. Sci., Part C,* no. 26 (1969), 99.

86. LeGrand, D.G. and Gaines, G.L., Jr., "Surface Activity of Block Copolymers of Dimethylsiloxane and Bisphenol-A Carbonate in Polycarbonate," *Polym. Prepr., Amer. Chem. Soc., Div. Polym. Chem.,* 11, no. 2 (1970), 442.

STRUCTURE AND MORPHOLOGY

MODERATOR: FRASER P. PRICE
Polymer Research Institute
University of Massachusetts
Amherst, Massachusetts

5. Morphology of Block Copolymers and Its Consequences

MICHAEL J. FOLKES AND ANDREW KELLER

University of Bristol
Bristol, U.K.

ABSTRACT

In block copolymers, microphase segregation of the individual blocks occurs, often leading to periodic structures of remarkable regularity. Indeed, it is the size, shape, and arrangement of the dispersed phase that must be determined before a thorough appraisal of physical properties can be attempted. The present article begins with a critical survey of some of the techniques employed for such structure studies and illustrates how specialized methods have developed to suit the particular sample requirements. Examples are given of the typical experimental results obtained and their relationship to molecular parameters. Some brief comments are made concerning the implications of current theories of microphase segregation and their comparison with present experimental data. Finally, some recent work concerning the behavior of macroscopic "single crystals" of block copolymers is discussed, in which precise correlations between morphology and properties can be achieved.

The present paper is a condensed review of the subject defined by the title. The first and major part is a general survey, while the last sections are primarily concerned with our own work at Bristol. We feel that our own work is a natural extension of what has been done in the field before us. It also represents some of the latest developments and thus has its place in the survey on its own right. The review is kept brief because a fuller version of it had already been submitted before the present article was commissioned. The reader is referred to this more extensive review [1] for further particulars, which also includes a reasonably complete bibliography (only a few sample references are given in here). Further, we took cognizance of the fact that some aspects of the field are the subjects of other separate presentations in the same issue. To avoid undue repetition we have curtailed the review portion accordingly.

General Review

In practice a combination of properties exhibited by different individual polymers is often desirable. However, it is intrinsic to polymeric systems that they are mutually incompatible and thus will defeat attempts at intimate mixing; in fact, the different constituents will segregate. Nevertheless, if a whole chain molecule of one species is linked chemically to another, as is the case in block copolymers, this segregation will be confined to submicroscopic domains, and in this way the desired combination of properties can still be achieved. The particular concern here is copolymer molecules comprising two or three blocks only, e.g., systems such as polystyrene–polybutadiene (S–B) or three-block versions of the same constituents such as S–B–S or B–S–B. The polystyrene–polybutadiene system is a particularly important example, as under ambient temperature one of the microphases (S) is glassy while the other (B) is rubbery, a combination that forms the basis of thermoelastomers. In these materials the microphase separation can often lead to surprisingly regular arrangements giving rise to a fully periodic structure, or macrolattice. (It is to be noted that such a lattice does not necessarily involve crystallinity on the molecular level; e.g., in the polystyrene–polybutadiene system both microphases are amorphous in themselves.) Such periodic structures formed by segregated microphases are the subject of the present article.

Periodic arrangements of microphases were recognized in earlier work on soap gels (e.g., [2,3]). It was recognized that among other more complicated possibilities, the soap molecules could take up lamellar, cylindrical, and spherical states of aggregation, which themselves give rise to a local mesomorphic structure, by forming one-, two-, or three-dimensionally periodic sequences. These very same forms and their modes of arrangement have subsequently been recognized in block copolymers, shown schematically in Figure 1. The conclusions in the soap work were reached by low-angle X-ray scattering. Consequently, when

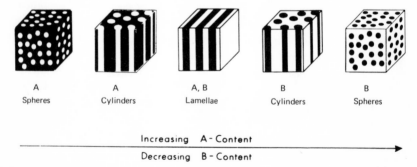

| A | A | A, B | B | B |
| Spheres | Cylinders | Lamellae | Cylinders | Spheres |

Increasing   A - Content
⟶
Decreasing   B - Content

Figure 1. Schematic representation of the dependence of block copolymer morphology on the volume fraction of the blocks. A and B denote the two chemical species forming the molecule, without reference to a particular system (Molau [29]).

block copolymers came into the forefront, little more was required than to apply the X-ray procedures already adopted in the study of soaps. Of course, low-angle X-ray diffraction, by its averaging nature, can be used to give only the over-all characteristics of the system under study. On the other hand, both rigid and swollen samples can be studied directly, without resource to any of the special experimental techniques required for electron microscopy, as subsequently described.

X-ray Scattering

The application of X-ray diffraction to this subject relies on the existence of one or more well-defined reflections at low angles. This implies the presence of an underlying periodic structure on a scale of the order of a few hundred Å. With the exception of some recent developments, in all the early studies on soaps and in all subsequent studies on block copolymers, this periodicity was randomly oriented over the cross-sectional area of the X-ray beam used, giving rise to Debye–Scherrer rings. From such a diffraction pattern the following information may be extracted: (1) periodicity of microphases—from ring diameter and Bragg's law; (2) geometry and type of macrolattice—from ratio of spacings of consecutive orders; and (3) size of the microphases—from relative intensities of the reflections or, alternatively, from a knowledge of the periodicity of the microphases and the stoichiometry of the sample.

Using this method of approach, Luzzati et al. [3] established the following geometries: (1) cubic lattice of spheres; (2) two-dimensional hexagonal lattice of cylinders; and (3) regularly repeating parallel lamellar sequence.

That the same geometries can exist for block copolymers has gradually emerged from both X-ray and electron-microscopic studies but was established by the extensive work of Skoulios and collaborators (e.g., [4]). These studies of the effects of preferential swelling agents for one of the phases of such systems as polystyrene–polyoxyethylene and polystyrene–polyvinylpyridine nicely illustrate the flexibility of the X-ray-diffraction technique as applied to block copolymers.

Everything said so far refers to random diffraction patterns giving rise to rings. Under special circumstances the full symmetry of the macrolattice may be displayed if the spatial extent of the microphase ordering is retained over areas at least as large as that encompassed by the X-ray beam and through the thickness of the sample used. This can result, e.g., from the particular processing conditions employed during the preparation of the sample [5]. More is said on this subject later.

Electron Microscopy

The confirmation of microphase ordering in block copolymers by using electron microscopy requires special sample preparation techniques, and this particular point, more than any other, determined the subsequent development of the subject. Indeed, here is a situation unique to structure studies, where direct observation of the structures and the analysis of the diffraction effects they produce can be closely correlated.

In quite general terms, there are at least three requirements that must be taken into account during sample preparation for electron microscopy.

1. With primarily a two-phase system, sufficient contrast must be created between the phases. This is accomplished by the now familiar technique due to Kato [6], in which the sample is stained in osmium tetroxide vapor. The rubbery phase only, becomes stained and appears black when viewed by transmission.

2. The sample must be sufficiently thin (10^2-10^3 Å) for the structure to be observed under transmission microscopy. This can be achieved either by solvent casting from a solution of the copolymer to yield a film of the required thickness or by sectioning from a thicker film or from the massive material.

3. The sample must be rigid if sectioning is to be carried out. For diluent-free samples, cryo-ultramicrotomy can be used [7,8]. Samples containing a diluent cannot be sectioned unless the diluent can subsequently be polymerized to preserve the structure during sectioning [9].

Requirements 2 and 3 represent the broad divisions into which individual works fall. In the first place, regular arrangements of microphases have indeed been observed (e.g., [10,11]). It has been found that in solvent-cast films the regularity is strongly influenced by the nature of the solvent and its rate of evaporation [11,12]. Thus, it appears that the slower the rate of evaporation of the solvent, the better the resulting microphase ordering. This suggests that structures formed in this way may not be in equilibrium, and this must accordingly be taken into account if comparisons with theory are made.

Further, it was customary in the early electron-microscopic studies to observe structures in only one projection through the film, usually normal to the film plane [10,12-14]. Conclusions drawn from a single observation can sometimes be ambiguous; e.g., circular dots can be interpreted as cross-sections of either spheres or cylinders. Although a note of caution is therefore needed in this method of approach, these studies very successfully set the scene for later work, in which the overall quality of the electron micrographs improved due to advances in techniques of staining and sample preparation.

Slightly later works tried to overcome this ambiguity due to seeing structures in one projection only. Lewis and Price [15] have calculated projected areas of the dispersed phase and have compared them with the known sample stoichiometry. Matsuo [11] was one of the earliest workers to recognize the advantages of bidirectional sectioning. His investigations were centered on a series of solvent-cast S-B, B-S-B, S-B-S, and S-B-S-B films. After hardening with osmium treatment, sections were cut parallel and perpendicular to the film surface. An impressive identification of lamellae is shown (Figure 2). The film surface in this

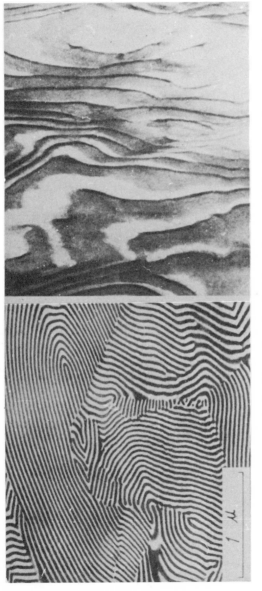

Figure 2. Electron micrograph of a thin section of an S–B–S copolymer containing 40% polystyrene and cast from cyclohexane solution. Sections cut normal (left) and parallel (right) to the film surface showing lamellar morphology (Matsuo [11]). Here, as in all subsequent electron micrographs, the white regions correspond to the polystyrene phase, and the black regions to the osmium-stained rubbery matrix.

case appeared to have imposed an orientation on the underlying textures, which allowed a distinction between spherical, cylindrical, and lamellar morphologies.

Hendus, Illers, and Ropte [13] were among the first to vary systematically the block ratio and sizes in a series of S–B–S solvent-cast or compression-molded films. Although the regularity of the phase separation was not as clear as in still later works, certain general conclusions were reached. Thus, the block component that was in smaller proportion was always the dispersed phase. The molecular weight for the same block ratio determined the scale of the structure.

These conclusions were confirmed and extended by Kämpf, Hoffmann, and Krömer in perhaps the most comprehensive work in this field (e.g., [8]). Most of their experimental work is concerned with B–S copolymers of varying block ratios and molecular weights. The electron micrographs (obtained from sections of solvent-cast films) revealed long-range order, which could extend over large areas for annealed specimens. These areas appeared as grains of crystallinity (Figure 3) sometimes extending over exceptionally large areas, revealing surprising uniformity (Figure 4). By covering a wide range of block ratios, the following structure sequence was established:

0– 25% S	S spheres in a B matrix
25– 40% S	S cylinders in a B matrix
40– 60% S	lamellar structures
60– 85% S	B cylinders in an S matrix
85–100% S	B spheres in an S matrix

This trend with block ratio was at least in part suggested by the other works and is implied in Figure 1. The samples used by Kämpf, Hoffmann, and Krömer were also studied by low-angle X-ray diffraction and light scattering. Excellent agreement was found for the dimensions and separation of the microphases using the different techniques.

All of this discussion presupposes that the original material is free of diluent. However, it is known that the presence of a preferential solvent can induce morphological changes in the dispersed phase. Sectioning of such material can now be achieved by using a swelling agent that is polymerizable. Douy and Gallot [9] employ such a method for the system S–B plus styrene. They obtained clear visual evidence for hexagonal, lamellar and inverse hexagonal structures for the S–B system. If styrene is added, e.g., to the lamellar system, it is found that the parallelism of the lamellae is first impaired and later, for higher solvent contents, completely removed.

Theory

The results discussed represent only a glimpse at the wealth of data from a large variety of different copolymer systems. It is encouraging to discover, how-

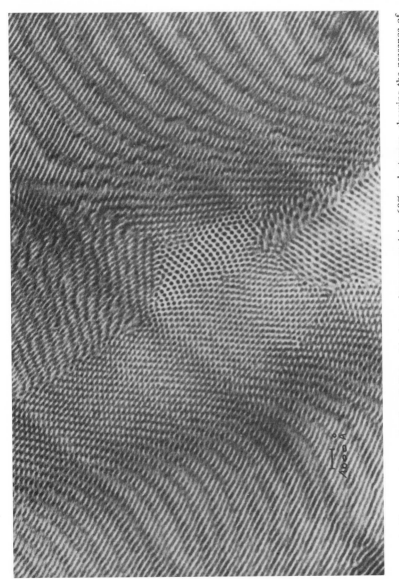

Figure 3. Electron micrograph of an S-B two-block copolymer containing 68% polystyrene, showing the presence of "grain" boundaries separating regions of microphase ordering (Kämpf, Hoffman, and Krömer [8]).

Figure 4. Electron micrograph of an S–B two-block copolymer containing 68% polystyrene, showing long-range microphase ordering (Kämpf, Hoffmann, and Krömer [8]).

ever, that the general trends revealed experimentally are in good agreement with current theoretical work concerned with the thermodynamics of phase separation in block copolymers (e.g., [16,17]).

Certain predictions resulting from these theories have been made.

1. *Influence of block-molecular-weight ratio.*   Theory suggests that the dominant factor influencing domain geometry is the volume fraction of the

phases. It supports the experimentally observed sequence of changes, spheres–cylinders–lamellae, as the proportion of the dispersed phase increases.

2. *Block molecular weight.* Block molecular weight primarily influences the scale of the microphase ordering (i.e., dimensions and separation of the phases). The work of Krömer et al. [16] has made it possible to obtain an experimental dependence according to which the diameters of spheres or cylinders and thickness of lamellae closely follow a square-root dependence on molecular weight, which is in good agreement with theoretical expectations.

3. *Molecular orientation in the phases.* According to theory, as a consequence of the uniform filling of space by the chains in both phases, the end-to-end chain separations may be different from those corresponding to a random coil; the effect, however, is expected to be slight.

4. *Interfacial region.* The magnitude and properties of the region forming the interface between the dispersed phase and the surrounding matrix also feature in many of the theories of phase segregation [18,19]. Experimental evidence that such regions indeed exist, with dimensions (often representing a significant proportion of the total dispersed-phase dimension) as predicted by theory, is very scarce. Direct inspection of electron micrographs certainly does not indicate any detectable interfacial zone that differs from the pure phases.

5. *Effect of temperature on domain dimensions.* Some of the theories predict a significant temperature dependence for the domain size [20]. However, low-angle X-ray measurements performed by Grosius et al. [21] and by ourselves at Bristol [22] reveal practically no change in domain parameters at least up to temperatures of $210°C$.

Macroscopic Single Crystals of Segregated Microphases

It was stated earlier in this review that under certain circumstances, samples exhibiting a uniformly-oriented macrolattice over macroscopic sample volumes can be obtained. The formation of such samples can be promoted by extrusion such as may occur during many fabrication processes in practical applications. Conditions leading to reproducible sample preparation have been recently established [5].

Samples

Most of the work concentrated on one particular S–B–S copolymer (Kraton 102, 25 percent S). As shown below, this material exhibited a cylindrical morphology. Recent studies have extended this work to other systems also, including an S-B-S sample displaying a lamellar structure [23].

Structure

X-ray diffraction. The detection of a regular macrolattice over macroscopic sample volumes relies primarily on low-angle X-ray-diffraction evidence. It was first observed in extruded and subsequently heat-treated plugs of Kraton 102 in a way exemplified by Figure 5 [24]. It can be seen that the diffraction pattern consists of isolated reflections such as characterize hexagonal single crystals. The implications are self-evident: there is a regular array of cylinders (which in this case consist of polystyrene); Figure 5(a) shows the reciprocal lattice viewed along the cylinder axis. The hexagonal lattice periodicity is 300Å; the cylinder diameter, as assessed both from the block composition and independently from the intensity of the reflections, is 150Å. Thus, the closest cylinder separation distance is also 150Å. The fact that the reflections are confined to one layer line in Figure 5(b) means that the cylinders are very long compared to their diameter.

As already stated, regular alternation of lamellae extending over macroscopic sample volumes has been achieved recently [23]. Evidence for this again relies primarily on X-ray diffraction used in the manner just described for the cylindrical structures.

Electron microscopy. In possession of such macroscopic single-crystal samples, one can carry out electron microscopy in a systematic manner through sectioning along preselected directions. Sections were cut perpendicular and parallel to the hexagonal axis (extrusion direction) on specimens cooled by liquid air (Figures 6 and 7) [7]. The end-on and side-on view of the cylinders, and the lattice they give rise to, could thus be directly verified as being in reasonable quantitative agreement with the diffraction patterns.

Properties

The possession of macroscopic single crystals of segregated microphases made possible the examination of various properties expected to be associated with the previously established anisotropic structure.

Birefringence. Samples of single-crystal structures of Kraton 102 (cylindrical morphology) displayed an optically uniaxial character, with the plug axis as the optic axis. The measured birefringence could be quantitatively attributed to form birefringence due to isotropic S cylinders within an isotropic B matrix. This implies that the molecular orientation is essentially random within both phases [25].

Infrared dichroism. The degree of molecular orientation within each phase could be determined by infrared dichroism by the choice of suitable dichroic absorption bands for each component. No molecular orientation could be detected in either phase within our experimental sensitivity, corroborating the previous birefringence result [26].

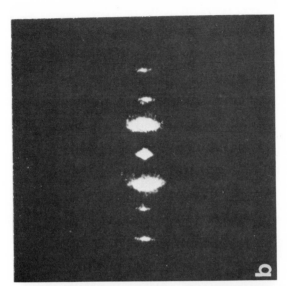

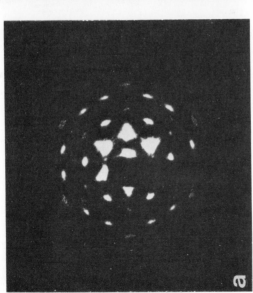

Figure 5. Low-angle X-ray diffraction patterns from a "single-crystal" sample of extruded and annealed S–B–S copolymer (Kraton 102): (a) beam parallel to the extrusion direction (plug axis); (b) beam perpendicular to the plug axis, which is vertical (based on Keller, Pedemonte, and Willmouth [24]).

Figure 6. Ultramicrotome section cut perpendicular to the extrusion direction of a "single-crystal" sample of extruded and annealed S–B–S copolymer (Kraton 102) (Dlugosz, Keller, and Pedemonte [7]).

Figure 7. Ultramicrotome section cut parallel to the extrusion direction of a "single-crystal" sample of extruded and annealed Kraton 102, showing striated structure (Dlugosz, Keller, and Pedemonte [7]).

Mechanical properties. The mechanical anisotropy is expected to be exceptionally large when one considers the samples as consisting of parallel glass fibers embedded in a rubbery matrix whose modulus is lower than that of the fibers by several orders of magnitude. Clearly one expects the material to be very much stiffer in the fiber direction.

Mechanical behavior was examined by cutting test samples from the extruded plug, with their long dimension making different angles θ with respect to the plug axis [25]. The stress–strain curves and corresponding Young's modulus E_θ were determined as a function of θ. Figure 8 shows stress–strain curves for some of the most important angles θ, while Figure 9 presents the corresponding E_θ values. The very significantly larger stiffness along the cylinder direction ($\theta = 0$) as compared with that for larger θ values is immediately obvious. Thus, these "single-crystal" samples behave as a glass and as a rubber in two mutually perpendicular directions, respectively. Even the slight minimum in E_θ at 55° is a predictable consequence of the constraint imposed by the cylinders on the matrix.

There is complete quantitative consistency in the variation of E_θ with θ from symmetry considerations for a transversely isotropic uniaxial system (dashed

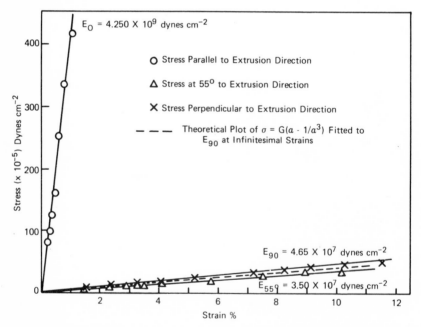

Figure 8. Stress–strain curves for some samples of a macroscopic "single crystal" of Kraton 102. E_θ is the value of Young's modulus measured at angle θ to the extrusion direction, and α is the extension ratio (Folkes and Keller [25]).

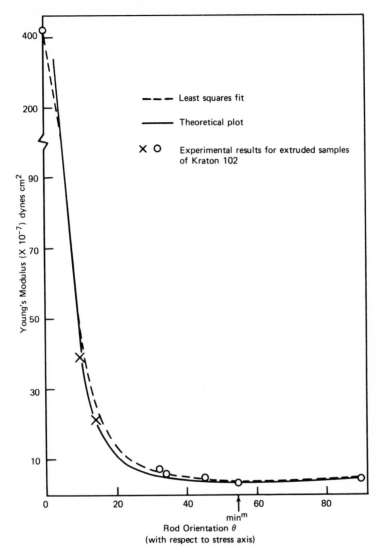

Figure 9. Variation of Young's modulus with orientation angle θ for some samples of a macroscopic "single crystal" of Kraton 102 (Keller et al. [22]; based on Folkes and Keller [25]).

line, Figure 9). Further, even the simplest series/parallel model closely matches the experiment except for the smallest θ values (solid line).

Clearly, with the aid of these macroscopic "single crystals," any other property can be studied in an analogous manner. Thus, we are currently investigating the swelling behavior that displays strikingly anisotropic effects, largely in the manner anticipated [22]. Detailed studies of properties of systems other than cylindrical are also in progress [27]. It is clear that the exploration of the properties of "single-crystal" samples is a prerequisite for the understanding of properties of samples prepared under more general conditions. Here more random polycrystalline textures are expected, where the relation between the properties of the individual microphases and aggregate properties are not so readily definable.

The usefulness of the "single-crystal" samples extends beyond the subject of block copolymers per se. Thus, they form ideal models for composite materials for the following reasons: the mechanical properties of the components differ greatly (glass-rubber); the anisotropy is very pronounced; and chiefly because of the molecular continuity, there is perfect contact between the components. In fact, more sophisticated theories of fiber composites have already been tested on "single crystals" of Kraton 102 (cylinders), and valuable conclusions have been made about the validity of the theories and the finer details of the copolymer systems themselves [28].

References

1. Folkes, M.J. and Keller, A., "The Morphology of Regular Block Copolymers," in *Physics of Glassy Polymers*, R.N. Haward, ed., London: Applied Science Publishers Ltd. (1973).
2. Sadron, C., "Organized Copolymers," *Chem. Ind., Genie. Chim.*, 96 (1966) 507. (In French)
3. Luzzati, V., Mustacchi, H., Skoulios, A., and Husson, F., "La Structure des colloides d'association. I. Les Phases liquide-cristallines des systèmes amphiphide-eau," *Acta Cryst.*, 13 (1960), 660.
4. Skoulios, A. and Finaz, G., "La structure des colloides d'association. VII. Caractère amphipatique et phases mésomorphes des copolymères séquencés styrolene-oxyde d'ethylène," *J. Chim. Phys.*, 59 (1962), 473.
5. Folkes, M.J., Keller, A., and Scalisi, F.P., "An Extrusion Technique for the Preparation of Single-Crystals of Block Copolymers," *Kolloid Z.-Z. Polym.*, 251 (1973), 1.
6. Kato, K., "Osmium Tetroxide Fixation of Rubber Latices," *J. Polym Sci., Part B*, 4 (1966), 35.
7. Deugosz, J., Keller, A., and Pedemonte, E., "Electron Microscope Evidence of a Macroscopic Single-Crystal from a Three Block Copolymer," *Kolloid Z.-Z. Polym.*, 242 (1970), 1125.
8. Kämpf, G., Hoffmann, M., and Krömer, H., "Sichtbare ubermolekulare Strukturen und weitreichende Ordnungen in nichtkristallinen Blockcopolymeren," *Ber. Bunsenges., Phys. Chem.*, 74 (1970), 851.
9. Douy, A. and Gallot, B.R., "Study of Liquid-Crystalline Structures of Polystyrene-Polybutadiene Block Copolymers by Small Angle X-ray Scattering and Electron Microscopy," *Mol. Cryst. Liquid Cryst.*, 14 (1971), 191.
10. Fischer, E., "Structures in Styrene-Butadiene Block Copolymers," *J. Macromol. Sci., Chem.*, A2 (1968), 1285.
11. Matsuo, M., "Polymerography," *Jap. Plast.*, 2 (1968), 6.

12. Lewis, P.R. and Price, C., "Morphology of ABA Block Copolymers," *Nature (London)*, 223 (1969), 494.

13. Hendus, H., Illers, K., and Ropte, E., "Strukturuntersuchungen an Styrol-Butadien-Styrol-Blockpolymeren," *Kolloid Z.-Z. Polym.*, 216–217 (1967), 110.

14. LaFlair, R.T., "The Structure Morphology and Properties of Block Copolymers," *23d International Congress of Pure and Applied Chemistry*, Vol. 8, London: Butterworths (1971), 195.

15. Lewis, P.R. and Price, C., "The Morphology of $(Styrene)_x$ $(Butadiene)_y$ $(Styrene)_x$ Block Copolymers," *Polymer*, 12 (1971), 258.

16. Krömer, H., Hoffman, M., and Kämpf, G., "Die ubermolekulare Struktur von Blockcopolymeren als Folge einer Unrertraglichkeit der Sequenzen und der molekularen Konformation," *Ber. Bunsenges, Phys. Chem.*, 74 (1970), 859.

17. Meier, D.J., "Theory of Block Copolymers–Domain Formation in A–B Block Copolymers," *J. Polym. Sci., Part C*, no. 26 (1969), 81.

18. Meier, D.J., to be published.

19. Leary, D.F. and Williams, M.C., "Statistical Thermodynamics of A–B–A Block Copolymers," *J. Polym. Sci., Part B*, 8 (1970), 335.

20. Bianchi, U., Pedemonte, E., and Turturro, A., "Morphology of Styrene–Butadiene–Styrene Block Copolymers," *Polymer*, 11 (1970), 268.

21. Grosius, P., Gallot, Y., and Skoulios, A., "Influence de la temperature sur la structure et les paramètres geometriques des copolymères sequences polystyrene/polyvinylpyridine," *C. R. Acad. Sci.*, 270 (1970), 1381.

22. Keller, A., Dlugosz, J., Folkes, M.J., Pedemonte, E., Scalisi, F.P., and Willmouth, F.M., "Macroscopic Single Crystals of an S–B–S Three Block Copolymer," *J. Phys.* (Paris), 32 (1971), C5a–295.

23. Dlugosz, J., Folkes, M.J., and Keller, A., "A Macrolattice Based on a Lamellar Morphology in an S–B–S Copolymer," *J. Polym. Sci.*, 11 (1973), 929.

24. Keller, A., Pedemonte, E., and Willmouth, F.M., "A Macrolattice from Segregated Amorphous Phases of a Three Block Copolymer," *Nature (London)*, 225 (1970), 538, and *Kolloid Z.-Z. Polym.*, 238 (1970), 385.

25. Folkes, M.J. and Keller, A., "The Birefringence and Mechanical Properties of a Single Crystal from a Three-Block Copolymer," *Polymer*, 12 (1971), 222.

26. Folkes, M.J., Keller, A., and Scalisi, F.P., "A Test for Molecular Orientation in a Single Crystal of S–B–S Three-Block Copolymer by Infra-red Spectroscopy," *Polymer*, 12 (1971), 793.

27. Folkes, M.J. and Keller, A., to be published.

28. Arridge, R.G.C. and Folkes, M.J., "The Mechanical Properties of a Single Crystal of S–B–S Copolymer–A Novel Composite Material," *J. Phys. D*, 5 (1972), 344.

29. Molau, G.E., "Colloidal and Morphological Behavior of Block and Graft Copolymers," in *Block Polymers*, S.L. Aggarwal, ed., New York: Plenum Press (1970), 102.

6. Block Copolymer–Solvent Systems: Thermodynamics and Morphology

DALE J. MEIER

Midland Macromolecular Institute
Midland, Michigan

ABSTRACT

A theory is presented for the thermodynamic and morphological behavior of block copolymer–solvent systems. Among the subjects discussed are the following: (1) the dependence of microphase separation or domain formation on such variables as polymer concentration, molecular weight, and interaction parameters; (2) domain morphology as a function of solvent content; and (3) the interfacial free energy in block copolymers.

It is emphasized that the morphology of the solvent-free block copolymer that has been cast from solution is generally a nonequilibrium morphology. That morphology which first forms during the evaporation of a block copolymer solution remains, even though other morphological forms may have a lower free energy in the solvent-free state.

In his pioneering work on graft copolymers, Merrett [1] observed that the physical properties of cast films of a methyl methacrylate-natural rubber graft copolymer were greatly dependent upon the solvent system used in casting. If cast from a solvent preferential to the poly(methyl methacrylate) component, the resulting film properties would be similar to those of a pure poly(methyl methacrylate) film, i.e., hard and stiff; if cast from a solvent preferential to natural rubber, the film would be soft and elastomeric. These observations have been extended in the intervening years to many graft and block copolymer systems [2,3], and the phenomenon is now recognized as a general one for these classes of polymers.

A qualitative explanation for the effects observed was given by Merrett [1] in terms of the chains of one component being collapsed (by an unfavorable interaction with solvent) into a micellar arrangement that was stabilized in solution by chains of the solvated component. The solvated component became the matrix or continuous phase when solvent was evaporated, and the properties of the recovered polymer would predominately reflect those of the component in this continuous phase. He also pointed out that an intermediate mixed state of both components would be higher in free energy than either of the extreme segregated configurations, and thus a barrier existed to prevent the ready transformation of one configuration to another. This explained the stability of the configurations against interconversion, even though, in general, they were in nonequilibrium morphological states. Quasi-equilibrium states are now recognized to be a ubiquitous feature of block and graft copolymers.

Although these qualitative explanations by Merrett are undoubtedly valid, little has been done in the following fifteen years to develop a deeper understanding and theory of the phenomena. Kawai and co-workers [4] and this author [5,6] have briefly discussed some facets of the thermodynamics of and morphological features associated with microphase separation in block copolymer–solvent systems. However, these theories are incomplete and allow few quantitative predictions. In the present paper, a theory of block copolymer–solvent systems is developed, with particular attention given to establishing thermodynamic criteria for microphase separation (domain formation) and resulting morphology.

Model and Assumptions

In general, the material presented in this paper follows the model and methods established in earlier papers of this series [5,6] (henceforth referred to as M1 and M2). Attention is restricted to A–B- and A–B–A-type block copolymers and the following is assumed: (1) the chains obey random-flight statistics; (2) the parameter C relating unperturbed chain dimensions and molecular weight is the same for the A and B components, $C = (\sigma l^2)_o^{1/2}/M^{1/2}$; (3) the densities of all components are 1.0 gm/cm^3 (the last two assumptions are unnecessary, but convenient since they greatly simplify notation and algebraic manipulation without introducing significant errors); (4) only three fundamental morphological forms of domains are possible (spheres, cylinders, and lamellae), and those of spheres and cylinders are arranged on a regular hexagonal close-packed lattice; and (5) the Flory–Huggins equation can be used to evaluate the free energies of mixing of the various components.

Domain Size and Solvent Content

In M1 and M2 it was shown that the theoretical relationship between chain dimensions and domain dimensions could be determined for each domain shape by invoking the physical requirement that space be uniformly filled with polymer segments. The derived relationships are of the form $D = k(\sigma l^2)^{1/2} = k\alpha(\sigma l^2)_o^{1/2} = kC\alpha M^{1/2}$, where D is a characteristic domain dimension, e.g., the radius of a sphere or cylinder or the thickness of a lamella; σ is the number of statistical elements of length l in the chain; α is a chain perturbation parameter (the ratio of perturbed to unperturbed chain dimensions, $\alpha^2 = (\sigma l^2)/(\sigma l^2)_o$); and k is a parameter that depends on domain shape and molecular architecture. The values of k that have been evaluated from this principle of uniform segment densities are as follows:

A-B polymers: Sphere $k = 1.33$

                Cylinder $k = 1.0$

                Lamella $k = 1.4$

A-B-A polymers: Sphere $k_B = 1.0$

                                $k_A = 1.33$

                Lamella $k_B = 1.2$

                                $k_A = 1.4$

Note the differences for the center- and end-block components of A-B-A-type polymers.

It is important to emphasize that the k's have been derived on the assumption that a uniform segment density is required. This is true whether or not solvent is also present (except, of course, at very dilute polymer concentrations where isolated molecules or aggregates of molecules exist—a subject not treated here). Hence, for the systems of interest here, the k's can be taken as constants. Since $(\sigma l^2)_o^{1/2}$ and the k's are then both constant for a given molecular weight and domain morphology, any change in domain dimensions, e.g., as a result of solvent addition, must reflect the change in the chain perturbation parameter α.

The chain perturbation parameters α_A and α_B are not independently variable [6] since the requirement for space-filling forces interrelationships between the ratio α_B/α_A and such factors as the relative number of A and B chains, domain morphology, molecular volumes, etc. As an illustration of this point, consider a lamellar domain system from an A-B block copolymer having block molecular weights of M_A and M_B. The relative quantity of solvent, if present, will be expressed as $q = v_s/v_A$, i.e., the ratio of the solvent volume to that of the A component. It is assumed that a fraction f of the solvent is in the A lamellae and the remaining fraction $(1 - f)$ is in the B lamellae. Neglecting the fraction of the system that constitutes the interfacial region of the domains, the number η_A of A chains per unit area of interface can be written as

$$\eta_A = \frac{T_A}{\bar{v}_A(1 + fq)} = \frac{k_A \alpha_A (\sigma_A l^2)_o}{\bar{v}_A(1 + fq)} = \frac{\overline{A}\rho_A k_A C\alpha_A M_A^{1/2}}{M_A(1 + fq)} = \frac{\overline{A}\rho_A k_A C\alpha_A}{M_A^{1/2}(1 + fq)}, \quad (1)$$

where T_A is the domain thickness, $\bar{v}_A$ is the molecular volume of an A chain, and $\overline{A}$ is Avogadro's number. A similar expression can be written for the B chains:

$$\eta_B = \frac{\overline{A}\rho_B k_B C\alpha_B}{M_B^{1/2}\left(1 + (1 - f)q\,\dfrac{M_A}{M_B}\right)}.$$

Since the number of A and B chains (or junctions) per unit area of the interface must be equal, we set $\eta_A = \eta_B$ and obtain the desired relationship between α_B and α_A:

$$\frac{\alpha_B}{\alpha_A} = \frac{k_A}{k_B}\left(\frac{M_B}{M_A}\right)^{1/2}\left(\frac{1 + (1 - f)q\,\dfrac{M_A}{M_B}}{1 + fq}\right) = \mu_l, \tag{2}$$

which for $q = 0$ reduces to the expression given earlier for A–B polymers [6]:

$$\frac{\alpha_B}{\alpha_A} = \left(\frac{M_B}{M_A}\right)^{1/2}.$$

Equation (2) shows that α_A and α_B are mutually dependent, and thus the size of a lamella cannot be related only to the molecular weight of the component in the lamella. The thickness of a lamella is a function of the molecular weights of both components (and, of course, solvent content and its distribution).

Similar considerations applied to cylindrical and spherical domains give the following relationships for the ratio α_B/α_A:

$$\text{Cylinders:} \quad \frac{\alpha_B}{\alpha_A} = \left[\frac{1.10\,\phi_d^{-1/2} - 1}{1.10\,\phi_d^{o\,-1/2} - 1}\right]\left(\frac{M_B}{M_A}\right)^{1/6} = \mu_C \tag{3}$$

$$\text{Spheres:} \quad \frac{\alpha_B}{\alpha_A} = \left[\frac{1.10\,\phi_d^{-1/3} - 1}{1.10(\phi_d^o)^{-1/3} - 1}\right] = \mu_s, \tag{4}$$

where ϕ_d is the volume fraction of the domain (assumed to be of the A component), e.g., $\phi_d = [\bar{v}_A(1 + fq)]/[\bar{v}_A(1 + q) + v_B]$, or, since we have assumed unit densities, we have $\phi_d = [M_A(1 + fq)]/[M_A(1 + q) + M_B]$, and (ϕ_d^o) is the domain volume fraction in the absence of solvent, i.e., $(\phi_d^o) = M_A/(M_A + M_B)$. The numerical factors in Equations (3) and (4) arise from relationships between inter-domain spacings (and related B chain dimensions), domain volume fractions, and domain radii (and related A chain dimensions). The problem of relating α_A and α_B of A–B–A polymers having B chains in the matrix is not considered here. In this case, α_A and α_B are not related from packing considerations alone since α_B can be changed by external stresses without a concomitant change in α_A. In the absence of solvent, it can be seen from Equations (2)–(4) that the mutual chain perturbations resulting from differences in block molecular weights increase in the order spheres $<$ cylinders $<$ lamellae, and that only in the case of spherical domains is the size only a function of the molecular weight of domain-forming component.

Thermodynamics

Interfacial Energy

In M1 and M2 it was assumed that the free energy associated with the domain interface could be characterized by an interfacial tension parameter, which was presumably accessible to measurement since implicitly it was thought to be the same as that for a simple mixture of A and B polymers. More recent work [8] on the theories of the homopolymer and the block copolymer interface has shown that this implicit assumption is incorrect. Not only is there no simple relation between the interfacial free energies in block copolymers and simple mixtures, but, more importantly, there is difficulty in even defining the interfacial tension γ at the block copolymer interface (e.g., if defined as $\gamma = (\partial G/\partial A)_{T,P}$, where G is the free energy and A the interfacial area, then at equilibrium the interfacial tension of the domain interface is zero. This result is a consequence of the fact that a dispersed phase can exist at equilibrium in block copolymers, in contrast to dispersed phases in simple mixtures, which always represent nonequilibrium states).

Although there are conceptual difficulties with the interfacial free energy, the problem does not exist for the interfacial energy; this quantity can be treated in a straightforward manner following the Cahn-Hilliard [9] formulation of the thermodynamics of nonuniform systems. Their treatment gives the interfacial energy ΔE of a planar surface of area A as

$$\Delta E = A \sum_{i>j} \frac{\chi_{ij} kT}{\bar{v}_i} \int_{\text{interface}} \left(\phi_i \phi_j + \frac{t^2}{6} \left(\frac{d\phi_i}{dx} \right)^2 \right) dx, \qquad (5)$$

where the interaction energy parameters are written in terms of the Flory χ parameter instead of the form used by Cahn and Hilliard; the ϕ's are segmental volume fractions, t is an "interaction range" parameter with dimensions of the order of segment sizes [8], and the other symbols have their usual meaning. For the present purposes, the second term within the integral is neglected by assuming that the gradient of composition within the interface is small. Although this assumption is not valid in general [8], it is acceptable here since the presence of solvents typically will reduce the interfacial gradient.

Placement and Segment Constraints

In M1 it was pointed out that the constraints placed on the position of block molecules in a domain array in which the centers of gravity of the molecules are no longer free and the constraints placed on the spacial position of segments in-

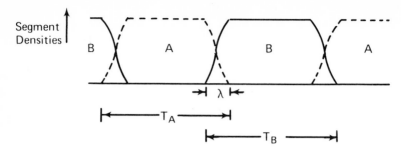

Figure 1. Cross-section of lamellar domains, showing segment density distribution.

creased the free energy over that of a random mixture. If the A–B junction of a block copolymer is restricted to being anywhere within the interfacial region of thickness λ, then the placement free energy [5] is given by

$$\Delta G_p = N_{AB} kT \ln A\lambda/V, \tag{6}$$

where N_{AB} is the number of molecules, and V is the total volume. ($A\lambda/V$ is, of course, merely the volume fraction of the interface.)

The free energy G_s associated with segment constraints is obtained (via the Boltzmann relation) from probability functions obtained as solutions to the diffusion equation [5], with boundary value conditions appropriate to the configurational constraints that keep segments segregated. Consider the segment density distribution of a lamellar domain system shown in Figure 1 as A lamellae of thickness T_A (measured to the outer edges of the interfacial region), B lamellae of thickness T_B, and an interfacial region of thickness λ. In order to obtain $G(s)$, we require the probability that molecules with the A–B junction within the interfacial region λ will have all A segments within the A domain volume and all B segments within the B domain volume. The probability $P_{AB}(T_A,T_B,x')$ that a chain with its junction at x' within λ (one interfacial boundary is at $x = 0$ and the other at $x = \lambda$) will have all segments in their proper segregated regions of space is

$$P_{AB}(T_A,T_B,x') = P_A(T_A,x') \cdot P_B(T_B,x') = \frac{16}{\pi^2} \sum_{m,n \text{ odd}} \sin \frac{m_\pi x'}{T_A} \cdot \frac{1}{mn}$$

$$\cdot \sin \frac{n\pi(\lambda - x')}{T_B} \exp \left\{ -\left(\frac{\pi^2}{6}\right)\left(\frac{\sigma_A l^2 m^2}{T_A^2} + \frac{\sigma_B l^2 n^2}{T_B^2}\right)\right\} . \tag{7}$$

The desired probability $P_{AB}(T_A,T_B,\lambda)$ that allows the junction segment to be anywhere within λ is obtained by integrating Equation (7) over x' and gives

$$P_{AB}(T_A,T_B,\lambda) = \frac{8}{3} \frac{\lambda^2}{T_A T_B} \exp - \frac{\pi^2}{6}, \tag{8}$$

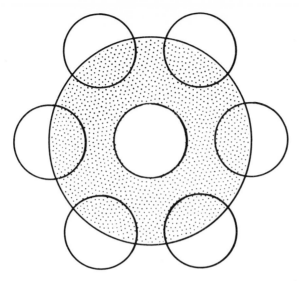

Figure 2. Constraint volume for matrix chains in spherical and cylindrical domains.

where λ/T_A and $\lambda/T_B \ll 1$ have been assumed and the earlier result for the appropriate value of k for $\sigma l^2/T^2$ has been used (only the first term in the summation over m and n is then required with this value of k).

The segment constraint free energy for cylindrical and spherical domains can be evaluated in a similar manner, except that the constraints on the B (matrix) chains require an element of approximation since the B environment is not isotropic, and "exact" solutions of the diffusion equation for the real domain geometry would be far too complex to be usable. However, the constraining space for the B chains can be made isotropic, and usable solutions then obtained by the approximation shown in Figure 2. The volume-constraining domains are dispersed in these systems, but their effect on the matrix chains can be approximated by smearing the domains such that they make a concentric constraining surface surrounding each domain. The matrix chains are constrained to stay within the shaded area shown in Figure 2. This approximation allows the B chains to be in regions of space that in actuality are excluded and, conversely, restricts the chains from regions of space that could be occupied. Thus, there are compensating errors introduced by this approximation. It is believed that with a suitable choice for the spacing of the outer constraining surface, the errors from this approximation should be minor. The following solutions to the diffusion equation are then obtained for the free energy of segment constraints:

Spheres: $G_s(s) = -NkT \ln \left[\dfrac{2}{3} \dfrac{(R' + R)}{(R' - R)} \dfrac{\lambda^2}{R^2} \exp - \dfrac{9\pi^2}{96} \left\{ 1 + \dfrac{(\sigma_B/\sigma_A)R^2}{(R' - R)^2} \right\} \right]$ (9a)

Cylinders: $G_s(c) = -NkT \ln \left[\dfrac{\pi}{3} \dfrac{\lambda^2}{R^2} M \exp - \dfrac{1}{6} \left\{ \beta_1^2 \dfrac{\sigma_B}{\sigma_A} \left(\dfrac{\alpha_B}{\alpha_A} \right)^2 \gamma^2 \right\} \right]$ (9b)

Lamellae: $G_s(l) = -NkT \ln \left[\dfrac{8}{3} \dfrac{\lambda^2}{T_A T_B} \exp - \dfrac{\pi^2}{6} \right]$, (9c)

where R is the domain radius, R' is the radius of the outer constraining surface (see below), and

$$M = \frac{J_o^2(\gamma R'/R)}{J_o^2(\gamma) - J_o^2\left(\dfrac{R'}{R} \gamma \right)} \left\{ [J_1(\gamma) N_o(\gamma) - N_1(\gamma) J_o(\gamma)] \right.$$
$$\left. - \frac{R'}{R} \gamma \left[J_1\left(\frac{R'}{R} \gamma \right) N_o(\gamma) - J_o(\gamma) N_1\left(\frac{R'}{R} \gamma \right) \right] \right\} , \quad (10)$$

where J_n and N_n are Bessel functions of the first and second kinds, respectfully, of order n, β_1 is the first root of $J_o(\beta) = 0$, and γ is the first root of

$$J_o(\gamma) N_o\left(\gamma \frac{R'}{R} \right) - N_o(\gamma) J_o\left(\gamma \frac{R'}{R} \right) = 0. \quad (11)$$

The radius R' of the outer constraining surface is taken, as shown in Figure 2, as the center-to-center distance of the domains;

$$R'/R = 1.905 \, \phi_D^{-1/2} \quad (12a)$$

for cylinders and

$$R'/R = 1.80 \, \phi_D^{-1/3} \quad (12b)$$

for spheres.

Free Energy of Mixing

The Flory–Huggins equation gives the change in the free energy of mixing:

$$\frac{\Delta G_M}{kT} = N_s^D \ln \phi_s^D + N_s^M \ln \phi_s^M + N_s^\lambda \ln \phi_s^\lambda + N_{AB} \phi_s^D \chi_{As} + N_{AB} \phi_s^M \chi_{Bs}$$
$$- N_{AB} \ln \phi_{AB}^o - N_s \ln \phi_s^o - N_{AB} \phi_B^o \chi_{AB} - N_{AB} \phi_s^o \chi_{Bs} - N_{AB} \phi_s^o \chi_{As}, \quad (13)$$

where the subscripts s and AB refer to solvent and block copolymer molecules, respectively, and the superscripts D, M, λ, o refer to domain, matrix, interface and original solution, respectively. Note that terms of the type $N_{AB}^D \ln \phi_A^D$ are not present since the centers of gravity of the AB molecules are fixed in the domain system, and terms of the type $N_{AB} \phi_s^\lambda \chi_{As}$ have been included as part of the interfacial energy.

Free Energy of Domain Formation from Solution

The free energy change ΔG in forming the domain system from a random solution of a block copolymer can be given by combining the interfacial energy from Equation (5), the placement and segment constraint free energy from Equations (6) and (9), the mixing free energy ΔG_M from Equation (13), and the chain perturbation ΔG_d free energy from

$$\frac{\Delta G_d}{N_{AB}kT} = \frac{1}{2}(\alpha_A^2 + \alpha_B^2 - 2\ln \alpha_A \alpha_B - 2) = \frac{1}{2}[\alpha_A^2(1+\mu^2) - 2\ln \alpha_A^2 \mu - 2], \quad (14)$$

where μ represents ratio α_B/α_A, as given in Equations (2-4). The free energy change is thus

$$\frac{\Delta G}{kT} = \sum_{i>j} A \chi_{ij} \int_0^\lambda \frac{\phi_i \phi_j}{\bar{v}_i} dx + N_{AB} \ln A \frac{\lambda}{V} + N_{AB} \ln \lambda^2 P_o + \frac{\Delta G_M}{kT}$$
$$+ \frac{N_{AB}}{2}[\alpha_A^2(1+\mu^2) - 2\ln \alpha_A^2 \mu - 2], \quad (15)$$

where the term for the segment constraints has been written as $\ln \lambda^2 P_o$ to show that λ^2 is a common factor for each domain shape. The term P_o represents the remaining portions of the expressions for the segment constraints in Equations (9a-c).

In the use of Equation (15), attention will be restricted to a few examples covering certain extremes in polymer-solvent interaction: (1) the solvent interacts equally with both components of the block copolymer, i.e., $\chi_{As}/\bar{v}_A = \chi_{Bs}/\bar{v}_B$; and (2) the solvent is highly preferential to one component, say the A component, which implies $\chi_{As}/\bar{v}_A \ll \chi_{Bs}/\bar{v}_B$.

Case 1. $\chi_{As}/\bar{v}_A = \chi_{Bs}/\bar{v}_B$. We may, without loss of generality, take $\chi_{As} = \chi_{Bs} = 0$ (solvent is a "good" solvent for the polymer species). Setting $q = N_s \bar{v}_s/N_{AB}\bar{v}_A$ and $r = \bar{v}_B/\bar{v}_A = \sigma_B/\sigma_A$, Equation (15) can be simplified to

$$\frac{\Delta G}{kT} = A \chi_{AB} \int_0^\lambda \frac{\phi_A(x)\phi_B(x)dx}{\bar{v}_A} - N_{AB} \ln A \frac{P_o \lambda^3}{V} - N_{AB} \ln \frac{1+r}{1+r+q}$$
$$- \frac{N_{AB}r\chi_{AB}}{1+r+q} + \frac{N_{AB}}{2}[\alpha_A^2(1+\mu^2) - 2\ln \alpha_A^2 \mu - 2]. \quad (16)$$

In order to evaluate the integral in Equation (16), which gives the interfacial energy, the functions $\phi_A(x)$ and $\phi_B(x)$ are required, giving the composition in the interface. A variational approach to obtaining these functions is possible [9], but it is somewhat pointless here since the final results are surprisingly independent of whatever reasonable function is used. A function that has the desired sigmoidal character (and is obtained from a theory [8] of the polymer-

polymer interface) is $\phi_i = \phi_i^o \sin^2 (\pi x/2\lambda)$, where ϕ_i^o is the change in composition across the interface. This functional form is adopted here and written as

$$\phi_A = \frac{1+r}{1+r+q} \sin^2 \frac{\pi x}{2\lambda} \quad \text{and} \quad \phi_B = \frac{1+r}{1+r+q} \cos^2 \frac{\pi x}{2\lambda}.$$

Note that for this case with $\chi_{As}/\bar{v}_A = \chi_{Bs}/\bar{v}_B$ it has been assumed that the solvent concentration is uniform throughout the system. This is not exactly true since the chain perturbation free energies contribute to the partial molal free energy of the solvent and affect the solvent distribution between the A and B phases. The effect is small, in general, and here it is ignored for simplicity. The interfacial energy then becomes

$$\frac{\Delta E}{kT} = A \chi_{AB} \int_0^\lambda \frac{\phi_A \phi_B}{\bar{v}_A} dx = \frac{A \chi_{AB}}{8 \bar{v}_A} \frac{(1+r)^2 \lambda}{(1+r+q)^2}, \tag{17}$$

and when it is substituted in Equation (16) and the resulting free energy minimized with respect to λ and α_A, we obtain

$$\frac{\partial(\Delta G)}{\partial \lambda} = \frac{A \chi_{AB}(1+r)^2}{8 \bar{v}_A(1+r+q)^2} - \frac{3N_{AB}}{\lambda} = 0; \quad \lambda_{min} = \frac{24 N_{AB} \bar{v}_A(1+r+q)^2}{A \chi_{AB}(1+r)^2} \tag{18}$$

$$\frac{\partial(\Delta G)}{\partial \alpha} = \alpha_A(1+\mu^2) - \frac{2}{\alpha_A} = 0; \quad \alpha_A^2(min) = \frac{2}{1+\mu^2}. \tag{19}$$

Substituting the value of λ_{min} in Equation (17) gives the surprisingly simple result for the interfacial energy $\Delta E = 3N_{AB}kT$, indicating that in this case the interfacial energy is independent of solvent content q, interaction parameter χ_{AB}, domain shape (this is not too surprising since it has been assumed that λ/R is small, and interfacial curvature has thus been ignored), and even the exact functional form of $\phi_A(x)$ and $\phi_B(x)$, as long as these functions give a direct proportionality between interfacial energy and interfacial thickness. Also, as seen from Equation (18), the interfacial volume λA is independent of domain shape. However, it must be noted that these very simple results are partially due to certain simplifying assumptions that have been made, e.g., neglect of the gradient term in Equation (5). Yet, a more rigorous treatment produces no drastic change in results; the interfacial energy remains approximately $3 N_{AB}kT$, but it is not constant and depends somewhat on domain shape.

The free energy associated with chain perturbations becomes with Equation (19)

$$\frac{\Delta G_d(min)}{N_{AB}kT} = \ln \frac{1+\mu^2}{2\mu},$$

where μ is given in Equations (2–4). The free energy change at minimum is

$$\frac{\Delta G(\text{min})}{N_{AB}kT} = 3 - \ln \frac{(24)^3 P_o V^2 (1 + r + q)^2}{A^2 \chi_{AB}^3 (1 + r)^5} - \frac{r \chi_{AB}}{1 + r + q} + \ln \frac{1 + \mu^2}{2\mu}. \tag{20}$$

Equation (20) may be used to establish critical concentrations for domain formation as a function of the interaction parameter χ_{AB}. For example, the following critical concentrations for a block copolymer having a block molecular-weight ratio (r) of 4 (with an assumed cylindrical domain shape) are obtained. The relevant parameters required to evaluate Equation (20) for cylinders, with $r = 4$, are $R'/R = 1.905$ $\phi_D^{-1/2} = 4.260$. In Equation (9b) for P_o, $M = 0.64$, $\gamma = 0.94$, and $\beta_1 = 2.405$. The interfacial area $A = 2V_D/R$, where V_D is the volume of the domains; and since $V_D = V/1 + r$, $A = 2V/R(1 + r)$. With these values, the free energy change becomes

$$\frac{\Delta G}{N_{AB}kT} = 2.616 - 2 \ln (5 + q) + 3 \ln \chi_{AB} - \frac{4\chi_{AB}}{5 + q}$$

or in terms of the volume fraction of polymer ϕ_p ($\phi_p = 1 + r/1 + r + q = 5/5 + q$),

$$\frac{\Delta G}{N_{AB}kT} = 2 \ln \phi_p + 3 \ln \chi_{AB} - \frac{4\phi_p \chi_{AB}}{5} - .603. \tag{21}$$

Presuming that domain formation occurs at $\Delta G = 0$, we find the following values for χ_{AB} are required for domain formation at various concentrations $\phi_p(\text{crit})$:

$\phi_p(\text{crit})$	χ_{AB}
1.0	5.9
0.1	112
0.01	1520

Although these values appear quite reasonable, further examination reveals some problems in that some of the assumptions made are not valid with these critical values. In particular, the assumption that the interfacial thickness is small with respect to domain dimensions is violated. For cylindrical domains being considered, the ratio of interfacial thickness to domain radius is given by Equation (18) as

$$\frac{\lambda}{R} = \frac{12(1 + r + q)}{\chi_{AB}(1 + r)}.$$

With the requirement here that λ/R be small, say $\lambda/R < 0.25$, we obtain

$$\chi_{AB} > \frac{48(1 + r + q)}{(1 + r)},$$

or in terms of volume fractions of polymer,

$$\chi_{AB} > \frac{48}{\phi_p} . \tag{22}$$

This condition on χ_{AB} is more restrictive (larger χ_{AB} values required for a given ϕ_p) than that based on $\Delta G = 0$ of Equation (21), and it is believed that Equation (22) represents a more realistic criterion for domain formation than does Equation (21). However, neither equation describes the onset condition in the formation of domains, which probably involves minor degrees of aggregation as a first step. During this first stage, the system would consist of some aggregated and segregated species and with considerable mixing of unlike segments throughout the system, rather than in a localized interfacial region as assumed in our model. The criterion here and its predicted critical values are thus for a later stage in the development of domain, where they are better developed.

It is also to be noted that Equation (22) shows that the critical concentration is inversely proportional to molecular weight (since χ_{AB} is proportional to the molecular weight of A).

There are no experimental data with which the present results can be directly compared. However, the data of Hoffman et al. [10] on styrene–butadiene block copolymers in benzene solutions offer some opportunity for comparison, although $\chi_{As}/\bar{v}_A$ and $\chi_{Bs}/\bar{v}_B$ are neither equal nor zero for this system, as was assumed in the theory, and the block-molecular-weight ratios r also differ somewhat (r's of 4 and 5). First, their experimental data show the predicted decrease of critical concentration with increasing molecular weight of the polystyrene block, but the data are too few and too scattered to determine whether the inverse proportionality of theory is obeyed. Second, for a polystyrene block of 10^4 molecular weight, the experimental data indicate a critical concentration of about 0.40 for a polymer with $r = 5$. From Equation (22), a critical concentration of about 0.48 is predicted, based on the following estimate for χ_{AB}. The "solubility parameter" difference $(\delta_1 - \delta_2)$ for polystyrene and polybutadiene is about 0.8 [7], which, with the relationship

$$\chi_{AB} = \frac{\bar{V}_A (\delta_1 - \delta_2)^2}{RT} ,$$

gives $\chi_{AB} = 106$ for $M_A = 10^4$.

The agreement is better than should be expected since the system is not exactly that treated by theory and because of the latitude possible in setting parameters to characterize the critical conditions (if $\lambda/R = 0.3$ had been arbitrarily used instead of $\lambda/R = 0.25$, the "agreement" between theory and experiment would have been almost perfect). In addition, the experimental "critical condition" is a function of the experimental method and can show considerable variation [10] depending upon the method used. This is a natural consequence of the microscopic size of the domains and the small refractive index difference of the phases at the critical state. However, even with these reservations, the fact that theory and experiment are not grossly different is considered encouraging.

Case 2. $\chi_{As}/\bar{v}_A \ll \chi_{Bs}/\bar{v}_B$. In the first example to be treated for this case, it is assumed that the differences in the interaction parameters are such that the solvent is completely in the A domains (it is recognized that this can never literally be the case in practice since it would require an infinite difference in the χ's). The free energy of domain systems is evaluated here as a function of solvent content, from which the equilibrium morphology can be established and comments can then be made on the observations of Merrett [1] and others [2] on the influence of preferential solvents on block copolymer properties.

In a discussion of domain morphologies as a function of solvent content, one need consider only relative free energies and thus can eliminate from consideration many of the terms in Equation (16) for the free energy since they are independent of domain shape. It was mentioned in an earlier section that as long as the interfacial energy is proportional to the thickness of the interfacial fraction, the interfacial energy is independent of solution parameters, and both the interfacial energy and interfacial fraction $A\lambda/V$ are independent of domain shape and hence need not be considered here. This is also assumed to be true for this case, so the relative free energy change from Equation (19) that need be considered is

$$\frac{\Delta G(\text{rel})}{N_{AB}kT} = -\ln P_o \lambda^2 + \ln \frac{1 + \mu^2}{2\mu}. \tag{23}$$

After substituting from Equation (9) for P_o, Equation (22) can be transformed to the following relative free energy changes, where q is again the ratio of the volume of solvent added to that of the A component, and r is the ratio of block molecular weights M_B/M_A (or volumes):

$$\text{Lamellae:} \quad \frac{\Delta G(\text{rel})}{N_{AB}kT} = -\ln\left\{\frac{6(1 + q + r)^2}{r(1 + q)}\exp-\frac{\pi^2}{6}\right\} + \ln\frac{1 + \mu_l^2}{2\mu_l}, \tag{24a}$$

where $\mu_l = \dfrac{r^{1/2}}{1 + q}$;

$$\text{Cylinders:} \quad \frac{\Delta G(\text{rel})}{N_{AB}kT} = -\ln\left\{\frac{3\pi}{4}M\exp-\frac{1}{6}(\beta_1^2 + r\mu_c^2\gamma^2) + \ln\frac{1 + \mu_c^2}{2\mu_c}\right\}, \tag{24b}$$

where M is given in Equation (10), β_1 in Equation (11), and μ_c in Equation (3);

$$\text{Spheres:} \quad \frac{\Delta G(\text{rel})}{N_{AB}kT} = -\ln\frac{2}{3}\left(\frac{R' + R}{R' - R}\right)\exp-\frac{9\pi^2}{96}\left\{1 + \frac{rR^2}{(R' - R)^2}\right\} + \ln\frac{1 + \mu_s^2}{2\mu_s}, \tag{24c}$$

where R'/R is given in Equation (12b), and μ_s in Equation (4).

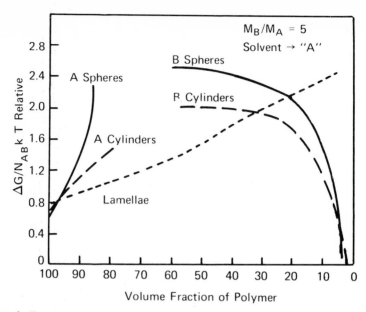

Figure 3. Free-energy and solvent-content relationships for domains from blocks of equal size.

Relative free energies calculated from Equations (24a–c) are shown in Figure 3 as a function of solvent content for a block copolymer for which $r = M_B/M_A = 5$. With no solvent present, the spherical domain system has the lowest free energy. However, when approximately 4 percent solvent is added (wholly to the minor A component), the lamellar domain then has the lowest free energy and remains lowest until the system is about 65% solvent, at which time the "inverted" B cylindrical domain system becomes lowest in free energy. Finally, when the solvent concentration is about 95 percent, a transformation of the domain system to B spheres is predicted. Thus, the addition of a preferential solvent to the minor component brings about profound morphological transformations and even causes an inversion of the role of the components as dispersed and matrix phases. It is obvious that, in principle, either component of a block copolymer could be forced to be the dispersed phase in the domain system by addition of a preferential solvent. A further example is shown in Figure 4 where the relative free energy of an A–B block copolymer, with $M_A = M_B$, is shown as a function of polymer concentration when a solvent preferential to A or to B is added. In this case, it is not assumed, as in the last example, that all of the added solvent goes into one component or the other, but a more realistic partitioning of 3/4 to one component and 1/4 to the other component is assumed. Figure 4 shows that the lamellar morphology has the lowest free energy in the

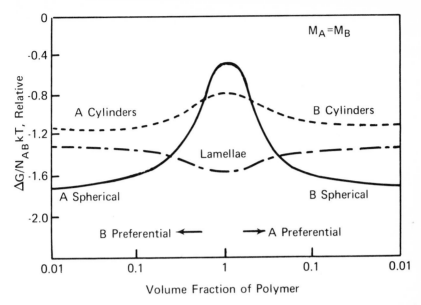

Figure 4. Free-energy and solvent-content relationships for different domain morphologies.

absence of solvent, but at concentrations below about 20 percent polymer, the spherical morphology is predicted to have the lowest free energy. Figure 4 also shows that the reasonable expectation that morphological transformations would occur in the sequence lamellar $\longrightarrow$ cylindrical $\longrightarrow$ spherical is incorrect. In this case the cylindrical morphology never has the lowest free energy; the transformation is directly from lamellar to spherical. This may account for some of the reported difficulties in obtaining block copolymer specimens with the cylindrical domain morphology [4].

Although domain morphologies have been discussed in terms of those having the lowest free energies, it is by no means assured that the system can attain that state with the lowest free energy. In fact, it is doubtful that domains having formed under one set of conditions could ever undergo a direct transformation to another shape in order to conform to a new lower-free-energy state. Merrett [1] foresaw the reasons for this many years ago, and he pointed out that the intermediate state in the direct transformation of domain morphologies is much higher in free energy than either end state; thus there is a large kinetic barrier to transformations. In molecular terms, domain transformations require the diffusional transport of chains of one component into and through an incompatible medium containing the other component. Thus, when block copolymers are recovered from solution, the solvent-free polymer can be expected to have the morphology of the domains first formed during the removal of solvent. As previously noted, this morphology is a function of block molecular weights and their ratios, of the preference of the solvent for one or the other components, and of the concentration at which domains form (and thus of the interaction parameters χ_{ij} of the system). These factors can be controlled and therefore offer the possibility of achieving any desired domain morphology for a desired end-use property.

References

1. Merrett, F.M., "Graft Polymers with Preset Molecular Configurations," *J. Polym. Sci.,* 24 (1957), 467.
2. Beecher, J.F., Marker, L., Bradford, R.D., and Aggarwal, S.L., "Morphology and Mechanical Behavior of Block Polymers," *J. Polym. Sci., Part C,* no. 26 (1969), 117.
3. Matsuo, M., "Structure and Properties of High-Molecular-Weight Compounds IV. Structures and Properties of Block Copolymers of Rubbers and Glassy Polymers," *Jap. Plast.,* 19, no. 3 (1968), 41.
4. Kawai, H., Soen, T., Inoue, T., Ono, T., and Uchida, T., "Domain Formation Mechanisms of Block and Graft Copolymers from Their Solutions," *Mem. Fac. Eng. Kyoto Univ.,* 33, pt. 4 (1971), 383.
5. Meier, D.J., "Theory of Block Copolymers. I. Domain Formation in A-B Block Copolymers," *J. Polym. Sci., Part C,* no. 26 (1969), 81.
6. Meier, D.J., "Theory of Morphology of Block Copolymers," *Polym. Prepr., Amer. Chem. Soc., Div. Polym. Chem.,* 11, no. 2 (1970), 400.
7. Brandrup, J. and Immergut, E.H., eds., *Polymer Handbook,* New York: Interscience (1966), 362.
8. Meier, D.J., to be published.
9. Cahn, J.W. and Hilliard, J.E., "Free Energy of a Nonuniform System. I. Interfacial Free Energy," *J. Chem. Phys.,* 28, no. 2 (1958), 258.
10. Hoffmann, H., Kampf, G., Kromer, H., and Pampus, G., "Kinetics of Aggregation and Dimensions of Supramolecular Structure in Noncrystalline Block Copolymers," in *Multicomponent Polymer Systems,* Washington, D.C.: American Chemical Society (1971), 351.

7. Organizational and Structural Problems in Block and Graft Copolymers

ANTHONY E. SKOULIOS

Centre de Recherches sur les Macromolécules
Strasbourg, France

ABSTRACT

Block copolymers are usually in the mesomorphic state. Their structure consists of blocks localized within distinct microdomains periodically distributed in space. Fairly free, the segregated blocks maintain their own habits: they crystallize or stay in the amorphous state, they behave as a glass or an an elastomer. The various types of structure that have been identified to date follow one another in a well-defined way, depending upon the length of the blocks and the amount of solvent present in the system. At equilibrium, interfaces between microdomains are far from diffuse. X-ray crystallography allows, at least in certain cases, evaluation of the partition coefficient of a given solvent between the segregated microdomains, and also study of the conformation of polymer chains as a function of temperature. Solubilization of homopolymers occurs when their molecular weight is lower than that of the corresponding blocks; it reveals a tendency for spontaneous elongation of polymer chains.

In the last decade low-angle X-ray diffraction techniques and electron microscopy have been successfully used to study the ability to organize and the structure of block copolymers. As a consequence, it is now well established that these polymeric materials are usually in the mesomorphic state. Because of their fundamental incompatibility, blocks segregate and, following their chemical nature, locate themselves within distinct microdomains. Intimately interdispersed, these microdomains are found to be related to one another in a very regular and periodic geometrical fashion. The perfectly organized systems thus produced are identical to the lyotropic liquid crystals observed in the binary mixtures of soaps and water [1,2].

The object of this paper is to deal with some particular problems associated with the overall architecture and geometry of block copolymers, aiming more specifically at the individual behavior of blocks and the way these are arranged

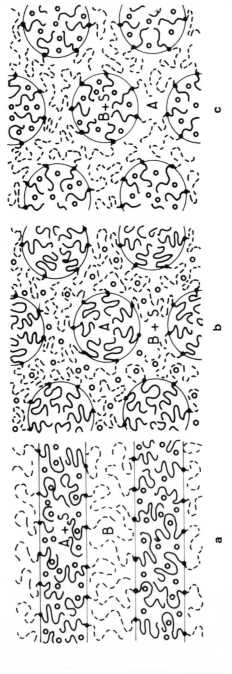

Figure 1. Schematic representation of (a) lamellar, (b) cylindrical, and (c) inverse cylindrical structures (small open circles represent solvent (S) molecules eventually present in the system).

within the microdomains. Not described is how, in 1960, it was found that block copolymers actually behave as liquid crystals [3], nor how low-angle X-ray diffraction was used to determine the structures they have been found so far to take up (Figure 1), that is, the lamellar [4], the cylindrical [4], the spherical [5,6], and the inverse cylindrical [6].

Most of the evidence supporting the ideas presented in this paper comes from X-ray diffraction studies carried out in the laboratory at Centre de Recherches sur les Macromolécules, and also in Dr. J. Terrisse's laboratory at Ecole d'Application des Hauts Polymères, Strasbourg.

Freedom and Crystallization of Blocks

The blocks being segregated and located within distinct microdomains rather large in size, there is nothing very peculiar about the fact that blocks have been found to maintain to an appreciable extent their intrinsic properties and their own habits, especially those related to short-range interactions. At any rate, this ability of block copolymers to keep the characteristics of their blocks has been widely used in the field of technology, particularly to produce physically vulcanized and moldable rubbers, high-impact plastics, and so on.

Among all the properties blocks can retain when forming part of a copolymer, the aptitude for crystallizing is considered here. As soon as the first liquid crystalline phases were identified in polystyrene–polyethyleneoxide block copolymers [4], it became apparent that, when free from solvent, polyethyleneoxide blocks, which are fundamentally stereoregular in nature and therefore capable of crystallizing, do indeed pass into the crystalline state. The structure of the system is then lamellar and results in the periodic and alternating pile-up of crystalline polyethyleneoxide and amorphous polystyrene layers (Figure 2). The polyethyleneoxide layers are very similar, if not identical, to those observed in semicrystalline homopolymers: polymer chains are oriented perpendicularly to the lamellae and are in the folded conformation. It seems, therefore, that the amorphous polystyrene blocks, which are chemically linked to the extremities of the polyethyleneoxide chains, do not hinder the latter from crystallizing in a very efficient way.

But how is the crystallization of a block affected when the other blocks are also capable of crystallizing? To answer this question, a series of polycaprolactone–polyethyleneoxide–polycaprolactone triblock copolymers have been studied by means of low-angle X-ray diffraction and dilatometry [7,8]. The conclusion reached is that, in this case also, the crystallizable blocks (here all of them) do in fact crystallize, producing a lamellar structure. The detailed architecture of the system is, however, slightly more complicated than that illustrated in Figure 2. It corresponds to a periodical pile-up of alternating

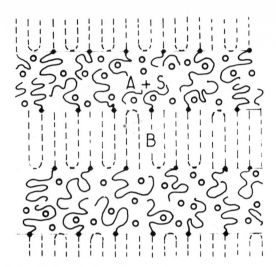

Figure 2. Schematic representation of a lamellar structure with crystalline (B) and swollen (A) blocks in solvent (S).

polyethyleneoxide and polycaprolactone layers, each of which is formed of a central sublayer of crystalline polymer inserted between two sublayers of amorphous material.

When the copolymer is rich in caprolactone, it behaves like a polycaprolactone homopolymer [7]. The thickness of its lamellae is independent of temperature in samples directly crystallized from the melt as well as in samples first crystallized at low temperatures and then annealed. When, on the contrary, the copolymer is rich in polyethyleneoxide, it behaves like a polyelthyleneoxide homopolymer; that is, the higher the temperature, the thicker its lamellae [7]. The main block thus imposes on the whole its own behavior and structure. This merely indicates that, since it crystallizes first, it freezes the entire structure and commands the crystallization of the other blocks.

When the copolymer is equally rich in caprolactone and ethyleneoxide [8], both blocks crystallize, but in an order dependent upon temperature. For the particular polymer that has been considered, polyethyleneoxide crystallizes first when temperature of crystallization from the melt is lower than 51°C, and last when temperature is higher than 52°C. In the intermediate temperature zone, both blocks crystallize with comparable rates of crystallization. As for the degree of crystallinity of the blocks, it has been found to be high (that is, of the same order of magnitude as for the corresponding homopolymers) only for those blocks that have crystallized first.

To summarize, with copolymers both blocks of which are crystallizable, blocks are still sufficiently free from one another to maintain their aptitude for crystallizing, but the overall structure of the system depends very much on circumstances. Whenever a block crystallizes first, it freezes the system in the structure that fits it better. Correspondingly, when it crystallizes last, it has to make the best of the structure that has already been established, and attains only low degrees of crystallinity.

Dependence of Structure Type upon Relative Lengths of Blocks and Solvent Content

Four types of structure have been identified to date for block copolymers—the lamellar, the cylindrical, the spherical, and the inverse cylindrical. Now the question arises of whether a given block copolymer system is capable of taking up all these structures and, if such is the case, of how the occurrence of a given structure depends on the relative lengths of the blocks and the solvent content of the system.

To answer this question, a series of polymethylmethacrylate–polyhexylmethacrylate two-block copolymers have been studied recently with the help of low-angle X-ray diffraction [6]. In this series, the molecular weight of the polyhexylmethacrylate block was a constant, $M_w = 8,900$, while the molecular weight of the polymethylmethacrylate block varied in the range from 2,400 to 43,100. The solvent used to swell the polymer was acetonitrile, which is a preferential solvent for the polymethylmethacrylate blocks.

Figure 3 illustrates the main results obtained from this work. It is worth noting, first, that, depending upon the respective lengths of the blocks and acetonitrile content of the gel, the system displays the whole range of structures described so far. It is also of interest to note that the overall width of the solvent concentration range in which the mixture is mesomorphic (and not an isotropic solution) decreases rapidly as the total molecular weight of the polymer decreases. As for the structures, it is perfectly clear that they follow one another in a well-defined way: as the molecular weight of the "soluble" polymethylmethacrylate block or the solvent content increases, the inverse cylindrical structure appears, followed by the lamellar, the cylindrical, and the spherical structures.

Intuitively, such behavior is easy to understand. Indeed, in the lamellar structure the interface between microdomains is a plane, and the lateral space offered to the blocks on either side of the interface is of comparable importance. It is not surprising, therefore, that this type of structure occurs in copolymers whose blocks are approximately of the same molar volume. Along this line of reasoning, it is also easy to understand why the structure is cylindrical when the "soluble" block is much longer than the "insoluble" one; this is merely due to the fact that on the convex side of the cylindrical interface the lateral space offered to the polymer chains is larger than that offered on the concave

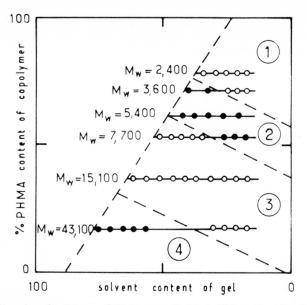

Figure 3. Domains of existence of (1) inverse cylindrical, (2) lamellar, (3) cylindrical, and (4) spherical structures for a series of polymethylmethacrylate–polyhexylmethacrylate two-block copolymers in the presence of acetonitrile used as a solvent (molecular weights indicated are those of polymethylmethacrylate blocks).

side. For the spherical structure to occur, the disproportion of the molar volumes of the "soluble" and "insoluble" blocks must be even more pronounced, since the spherical interface is doubly curved. The presence of a solvent in the system contributes to increasing the apparent volume of every "soluble" block; this is why the "soluble" blocks are generally located outside the structural elements, between the cylinders or the spheres, and also why, on increasing dilution, one first meets the lamellar structure, then the cylindrical, and lastly the spherical. However, when the "soluble" blocks are very short, their apparent volume is also very small, even in the swollen state. As a result, they cannot remain on the convex side of the interfaces, or even within a plane microdomain boundary; they therefore switch toward the interior of the structural elements, yielding the inverse cylindrical structure.

Solubilization of Homopolymers and Conformation of Chains

When the segregation of blocks and their ability to maintain individual properties had been established, the question arose of whether they could dissolve homopolymers and drag them within the microdomains where they themselves are located and, if such were the case, of how the overall structure of the system would be affected. To answer this question, a study has recently been undertaken, some preliminary results of which are now presented [9].

A particular mesomorphic block copolymer system has been considered, which is a 50 percent by weight mixture of a fairly monodisperse polystyrene–poly(vinyl-2 pyridine) two-block copolymer with octanol, which dissolves the pyridine blocks. The molecular weight was 9,500 for the polystyrene and 10,600 for the polyvinylpyridine blocks. The structures of the gel proved to consist of polystyrene cylinders 108 Å in diameter, imbedded in a matrix of polyvinylpyridine swollen in octanol.

First, the dissolving power of this gel was investigated with respect to highly monodisperse homopolystyrenes of different molecular weights ranging from 3,820 to 22,400. It was thus shown that low-molecular-weight homopolystyrenes (up to 9,500) are easily solubilized, leading to clear gels, while polystyrenes of molecular weight higher than 9,500 are incompatible with the mesophase, leading to opaque gels. This solubilization behavior has been confirmed by low-angle X-ray diffraction. Because homopolystyrene is in all cases incompatible with polyvinylpyridine and octanol and therefore can only locate itself if dissolved within the polystyrene microdomains of the block copolymer, this very sharp dependence of solubility upon molecular weight merely indicates that homopolystyrene has a direct means of comparing its molecular size to that of polystyrene blocks: if its molecular weight is equal to or smaller than that of the blocks, it dissolves; if its molecular weight is higher, it remains insoluble. Obviously this ability would be very poor if homopolystyrene were hidden in the very center of the structural elements (Figure 4a), away from the

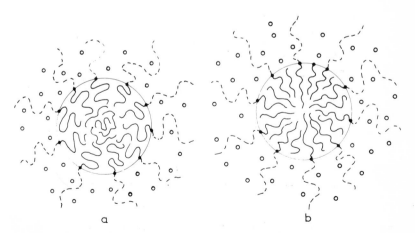

Figure 4. Schematic representation of solubilization: homopolymer is (a) located in the center of structural elements, or (b) inserted among chains of like blocks.

polystyrene blocks. The only model that seems satisfactory from this standpoint is illustrated in Figure 4b. The homopolymer chains, having the same conformation as the blocks, fit in a form of "palissade" and are inserted among the chains of their like blocks. Thus, if short, they are efficiently protected against the imbedding matrix; if long, in order to avoid their partial entry into the swollen polyvinylpyridine, which would be unfavorable from the viewpoint of interactions, they prefer to form a separate phase of pure homopolymer.

Second, the evolution of the structure was studied as a function of the homopolymer content of the system. The homopolystyrene used was of the same molecular weight as its like blocks. Figure 5 illustrates the results that were obtained. Up to concentrations of about 25 percent homopolymer, the gel is a homogeneous one-phase system, and above that limit the excess homopolymer precipitates into a separate phase. Within this concentration interval, the structural parameters change in a continuous way. The Bragg spacing d of the two-dimensional hexagonal array (spacing that is proportional to the distance between the axes of neighboring cylinders) and the radius R of the cylinders increase rapidly as concentration p increases (p = amount of homopolystyrene/ amount of polystyrene blocks in the gel). It is of interest to note that S_{cop}, which represents the mean lateral packing area of one "soluble" block at the interface, is little affected by the addition of homopolymer. Instead, S_{chain} decreases rapidly with increasing p: the packing of insoluble blocks becomes

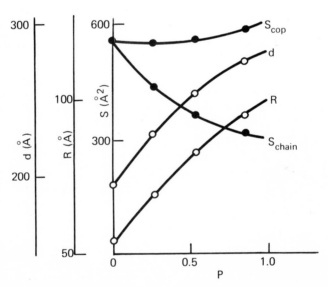

Figure 5. Variations of Bragg spacing d, of radius R of cylinders, and of lateral packing areas at the interface of "soluble" (S_{cop}) and "insoluble" (S_{chain}) blocks, as a function of p (relative amount of homopolymer to like blocks).

tighter because of their insertion among the homopolymer chains. The resulting elongation of polystyrene is important because it can attain values corresponding to half the total length of chains.

The tentative conclusion that could be drawn from this work is that polystyrene chains, despite their amorphous state, have not so chaotic a conformation as one might expect at first sight, but stretch out very readily without appreciable loss of stability of the system.

X-Ray Crystallography of Lamellar Structures

Before proceeding further, it is worth recalling briefly what has been learned about lamellar colloidal systems by means of X-ray diffraction. This particular structure has been selected for discussion here because it leads to mathematical expressions that are easy to handle and to use. All the conclusions drawn below can be simply extended to other types of colloidal structures, such as the cylindrical structure or the spherical one.

Let us consider a lamellar colloidal structure resulting from a periodical pile-up of perfectly alternating A and B layers (Figure 6). Reflection of X-rays

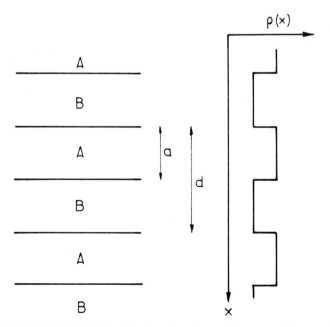

Figure 6. Schematic representation of a lamellar pile-up of layers A and B, and of electron density distribution along the normal to the layers.

by such a system occurs over a narrow angular range around the directions defined by Bragg's law [10]:

$$s = \frac{2 \sin \theta}{\lambda} = \frac{n}{d},$$ (1)

where 2θ is the angle between the diffracted and the direct beam, λ the wavelength of the X-ray radiation used, d the thickness of the unit lamella (that is, of an A and B layer superposed), and n an integer, called order of diffraction.

The intensity of the beam diffracted at a reciprocal spacing s is given by

$$I(s) = [F(s)]^2,$$ (2)

where $F(s)$ is the structure factor [10] of the system. This factor is directly related to the electron density distribution $\rho(x)$ along the normal to the layers (Figure 6):

$$F(s) = \int_{-d/2}^{+d/2} \rho(x) \exp(-2\pi i x s)\, dx.$$ (3)

In the particular case where the electron density is constant within each layer and the interface between layers infinitely sharp, the structure factor is

$$F(s) = \frac{\sin \pi a s}{\pi a s},$$ (4)

where a is the thickness of one of the layers, e.g., of the A layer.

If, now, the interfaces are not infinitely sharp, that is, if the electron density varies continuously from its constant value in the pure A layers to its value in the pure B layers, the structure factor and the intensity of the diffracted beams will be different. One simple way of representing this mathematically is to assume that the actual electron density distribution takes the form of a convolution product,

$$\rho(x) * g(x),$$ (5)

where $\rho(x)$ is the corresponding electron density with sharp interfaces, and $g(x)$ a distribution function defining the electron density change around the actual interfaces. By using Equation (3), it may be shown that the actual structure factor is

$$F(s) = \frac{\sin \pi a s}{\pi a s} \cdot f(s),$$ (6)

where $f(s)$ is the Fourier transform of $g(x)$. To simplify the calculations, it is convenient to take for $g(x)$ a form similar to that of the electron density itself (Figure 7). The width X of this distribution, then, represents the thickness of

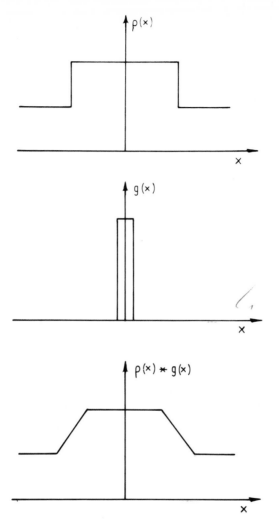

Figure 7. Schematic representation of electron density distribution $\rho(x)$ for sharp interfaces, of function $g(x)$, and of electron density distribution $\rho(x) * g(x)$ for diffuse interfaces.

the diffuse interface, and Equation (6) becomes

$$F(s) = \frac{\sin \pi as}{\pi as} \cdot \frac{\sin \pi Xs}{\pi Xs}. \tag{7}$$

This means that the actual structure factor is equal to the theoretical structure factor corresponding to sharp interfaces, modulated by just a sinusoidal term, the wavelength of which depends on the diffuseness of the interfaces.

TABLE I

Absent Orders of Diffraction as a Function of ϕ

ϕ	n
1/2	2,4,6, ...
1/3	3,6,9, ...
1/4	4,8,12, ...
2/5	5,10,15, ...
1/6	6,12,18, ...
3/7	7,14,21, ...

To use this mathematical treatment in the case of block copolymers, it is convenient to introduce a new parameter ϕ representing the volume fraction of A layers in the system

$$\phi = a/d. \tag{8}$$

Combination of Equations (1), (2), (7), and (8) leads to

$$I(n) = \left[\frac{\sin n\pi\phi}{n\pi\phi} \cdot \frac{\sin n\pi\phi(X/a)}{n\pi\phi(X/a)} \right]^2, \tag{9}$$

where $I(n)$ is the intensity of the nth order of diffraction for a lamellar system in which the volume fraction of A is ϕ, and the thickness of the diffuse interface between A and B layers is characterized by X/a.

To conclude this section, it is of interest to show how sensitive the dependence of $I(n)$ is on n and ϕ, especially in the case where the thickness of inter-

TABLE II

Ratio $I(n + 1)/I(n)$ of the Intensities of $n + 1$ and n Orders of Diffraction as a Function of n and ϕ

ϕ	$n = 9$	$n = 8$	$n = 7$	$n = 6$	$n = 5$
0.50	0	∞	0	∞	0
0.49	0.08	11.78	0.05	19.93	0.02
0.48	0.39	2.43	0.22	4.44	0.10
0.47	1.21	0.74	0.57	1.60	0.25
0.46	4.04	0.20	1.34	0.64	0.50
0.45	33.10	0.02	3.36	0.23	0.91
0.44	46.64	0.01	12.33	0.06	1.65
0.43	3.36	0.13	748.75	0.00	3.16
0.42	0.69	0.39	17.85	0.03	7.24
0.41	0.11	0.91	2.88	0.12	27.93
0.40	0	2.07	0.77	0.28	∞

faces is negligible. First, it is obvious that whenever $n\phi$ is an integer, the intensity is nil and the corresponding diffraction lines are absent from the diffraction pattern. Table I shows a few special cases of this type. For example, if $\phi = 2/5$, all the diffraction lines the order n of which is a multiple of 5 are systematically extinguished and invisible. But the value of ϕ not only controls the existence of diffraction lines but also affects the distribution of the intensities of the diffracted beams. Thus, minute changes in the values of ϕ can drastically affect the relative intensities, especially of the higher orders of diffraction. It is thus possible, just by comparing the intensities of the $n + 1$ and n orders of diffraction, to determine quite accurately the volume fraction ϕ of A and B layers in the system (Table II). For example, if the 6th, 7th, and 8th orders of diffraction are of comparable intensities $(I(7)/I(6) \sim I(8)/I(7) \sim 1)$, then ϕ has a value of approximately 0.465.

Sharpness of Interfaces

One of the questions that arise whenever one is trying to describe the structure of block copolymers is: How perfect is the segregation of the blocks and how sharp are the interfaces between microdomains?

In an attempt to answer this question, X-ray crystallography has been recently used to study the particular case of the lamellar structure of a two-block polystyrene–polyisoprene copolymer [11]. The polymer chosen had a total molecular weight of $M_w \sim 3.10^5$ and a chemical composition such that the volume fraction of each block was exactly $\phi = 0.50$. The Bragg spacing of the lamellae was $d \sim 900$ Å, and the thickness of every layer of $a \sim 450$ Å.

In perfect agreement with the theoretical predictions of Equation (9), the only diffraction lines present in the experimentally registered X-ray patterns were those for which n was odd. In addition, the intensities of the visible diffraction lines decreased monotonically with increasing n.

Let us now consider the overall distribution of intensities in order to deduce the thickness of the interfaces between microdomains. It is important to report here that X-ray patterns were found to contain up to the 15th order of diffraction and that, from a simple visual inspection of the diagrams, the intensity of the 15th order was approximately a hundred times weaker than that of the first order.

From Equation (9), and taking into account that $\phi = 0.50$, one finds

$$\frac{I(15)}{I(1)} = \frac{\sin^2\left(15\pi\frac{1}{2}\frac{X}{a}\right)}{15^4 \sin^2\left(\pi\frac{1}{2}\frac{X}{a}\right)}.$$

Considering then that $I(1) \sim 100\,I(15)$, one calculates for X/a the value of 0.026. This means that the thickness of the diffuse interfaces cannot exceed a value of 12 Å, which is not far from the actual size of monomeric units within the polymer. It must be stressed here that this value is in fact an upper limit for the thickness of the interfaces; indeed, if the lamellae composing the system are not strictly plane, but slightly undulating, the distribution of the intensities of the diffraction lines changes as though the interfaces were rough and diffuse.

Partition of Solvents Among Microdomains

Solvents added to organized block copolymers are usually assumed to be located preferentially or exclusively within those microdomains containing the blocks for which they show a better affinity. This description permits one to calculate the value of structural parameters such as size of microdomains or lateral packing areas of macromolecules at the interface [12]. However, such values are realistic only when the partition coefficient of the solvents among microdomains has been determined directly by experiment. The only parameter that can be evaluated without any assumption about the partition coefficient is the lateral packing area of macromolecules in the particular case of a lamellar structure. Thus, one has to be very careful while discussing the physical signif-icance of structural parameters, especially those related to the size of micro-domains; otherwise one is frequently led to erroneous interpretations and conclusions [13].

In order to determine the partition coefficient of a given solvent inside an organized copolymer, one often assumes that this coefficient is not very dif-ferent from what it is in the case of a mechanical mixture of two homopolymers identical to the copolymer blocks. One then measures its value just by analyzing the degree of swelling of the two homopolymers separated into two distinct and incompatible phases [12]. But this method is, of course, arbitrary and approx-imate. Application of the X-ray diffraction theory mentioned above allows a direct and better evaluation of this parameter.

Indeed, by analyzing, with the help of Equation (9), the distribution of the intensities of the diffracted X-ray beams, one has a means of evaluating directly the thickness of each layer forming the lamellar structure and of deducing there-from the degree of swelling of each block and the partition coefficient of any solvent.

This method has recently been successfully applied to the case of the lamellar structure of a polystyrene–polyisoprene two-block copolymer [11]. The molecular weight of each block was approximately 32,000. Table III sum-marizes the results obtained with a variety of solvents; the accuracy claimed for the values obtained is 10 percent. It can be seen that dodecane, which is known

TABLE III

Volume Fraction of Solvent Located Within
Polyisoprene Microdomains

Solvent	25°C	56°C
Dimethylformamide	0.10	0.10
Toluene	0.63	0.50
Ethylbenzene	0.70	0.60
Styrene	0.60	0.50
Cyclohexane	0.86	0.70
Pentylbenzene	0.85	0.75
Decylbenzene	1.00	0.80
Dodecane	1.00	1.00

to be an exclusive solvent for polyisoprene, does indeed locate itself within polyisoprene microdomains. On the other hand, it can also be seen that styrene, considered a preferential solvent of polystyrene [13], is in fact equally distributed between polystyrene and polyisoprene microdomains.

Temperature Dependence of Structural Parameters

It is well known that, because of thermal agitation of atoms and molecules, the lattice parameters of crystals increase upon heating, merely because of thermal expansion. However, in the case of organized colloids, which are liquid crystalline in character, the lattice parameters decrease with increasing temperature, despite the overall thermal expansion of the gel. Thus, with binary mixtures of soaps and water [2], the Bragg spacings of the mesophases and the size of paraffin microdomains decrease upon rising temperature, while the interfaces between paraffin and water regions expand laterally to an appreciable extent. How do organized block copolymers behave in this respect, and how does the polymer chain conformation depend on temperature?

In an initial attempt to answer this question, an X-ray diffraction study was carried out on a series of polystyrene-polyvinylpyridine block copolymers in the presence of a variety of solvents [14]. It was shown that repeat distances of the structural elements—lamellae, cylinders, and spheres—are almost insensitive to a change of temperature, indicating that the conformation of polymer chains, itself, is not very dependent upon thermal motion of atoms. It should be kept in mind, however, that these results were obtained on a particular polymer system containing large amounts of solvent and formed of copolymers the blocks of which were not true elastomers characterized by high flexibility of chains.

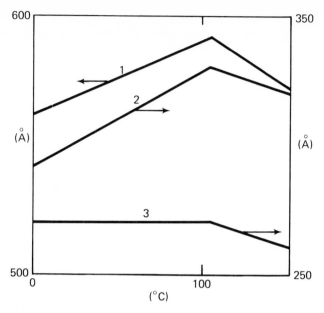

Figure 8. Variation of (1) Bragg spacing d and of layer thicknesses of (2) polyisoprene and (3) polystyrene blocks.

Complementary results have been obtained recently [11] in the absence of solvent and with a copolymer formed of highly flexible isoprene blocks and polystyrene blocks capable of entering the glassy state. The polymer used was a two-block copolymer, each block of which had a molecular weight of about 70,000. Containing more than 10 diffraction lines, X-ray patterns gave the Bragg spacing d of the lamellar structure with an accuracy of 1 percent. Figure 8 illustrates how d depends on temperature in the range from 0 to 150°C. It is worth noticing that up to about 105°C d increases with temperature, while beyond this limit it decreases rapidly. It is important to note also that this dependence of d on temperature is perfectly reversible, which means that the thickness of polystyrene and polyisoprene layers is related to an equilibrium conformation of polymer chains. Such an interpretation is strengthened by the fact that the glass transition temperature of polystyrene blocks (85°C), determined by dilatometry, is lower than the temperature where the maximum value of d was registered: mobility of polymer chains does not, apparently, affect the structural parameters of the system.

The block copolymer used had a chemical composition such that the volume fractions of blocks were about $\phi = 0.50$. As a consequence, the intensity of the X-ray diffraction lines was very sensitive to temperature (Figure 9). By using Equation (9) of the diffraction theory presented above, the thermal dependence

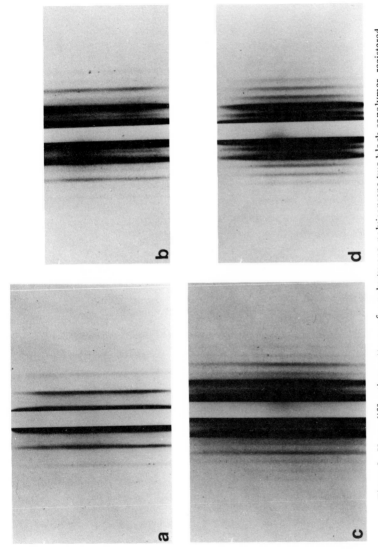

Figure 9. X-ray diffraction patterns of a polystyrene–polyisoprene two-block copolymer, registered at (a) −60, (b) 0, (c) 22, and (d) 65°C.

TABLE IV

Volume Fraction ϕ of Polyisoprene
as a Function of Temperature

Temperature (°C)	ϕ
–60	0.50
0	0.52
22	0.53
65	0.54
100	0.56
150	0.56

of ϕ was determined (Table IV), and the respective thicknesses of polystyrene and polyisoprene layers were calculated (Figure 8). In spite of the rather poor accuracy of this kind of evaluation, it is clear that up to about 105°C the polystyrene layers do not change in thickness, all the expansion of d being due solely to polyisoprene. Above 105°C, both polystyrene and polyisoprene layers decrease simultaneously in thickness, indicating that the volume expansion of the sample is entirely reflected in lateral spreading of layers.

Without prejudging conclusions to be drawn in the future, it is tempting to interpret the behavior of block copolymers described here as the result of conflicts between the stiffness of polystyrene and the flexibility of polyisoprene chains. Further investigations in this field should provide valuable information for a better understanding of the conformation of polymer chains.

Acknowledgment

The author is indebted to Dr. J. Terrisse for helpful and stimulating discussions on block copolymers and for many data presented in this paper.

References

1. Luzzati, V., Mustacchi, H., and Skoulios, A., "Structure of the Liquid–Crystal Phases of Some Soap and Water Systems," *Discuss. Faraday Soc.,* 25 (1958), 43.
2. Skoulios, A., "The Structure of Concentrated Aqueous Soap Solutions," *Advan. Colloid Interface Sci.,* 1 (1967), 79.
3. Skoulios, A., Finaz, G., and Parrod, J., "Obtention de gels mésomorphes dans les mélanges de copolymères séquencés styrolène-oxyde d'éthylène avec différents solvants," *C.R. Acad. Sci.,* 251 (1960), 739.

4. Skoulios, A. and Finaz, G., "La structure des colloides d'association. VII. Caractère amphipathique et phases mésomorphes des copolymères séquencés styrolène-oxyde d'éthylène," *J. Chim. Phys.,* 59 (1962), 473.

5. Tsouladze, G. and Skoulios, A., "La structure des colloides d'association. IX. Caractère amphipathique et phases mésomorphes des copolymères séquencés oxyde de propylène-oxyde d'éthylène,' *J. Chim. Phys.,* 60 (1963), 626.

6. Ailhaud, H., Gallot, Y., and Skoulios, A., "Structure of Mesomorphic Phases Observed with Block Copolymers of Different Alkyl Methacrylates," *Kolloid Z.-Z. Polym.,* 248 (1971), 889.

7. Perret, R. and Skoulios, A., "Etude de la cristallisation des copolymères triséquencés polycaprolactone-polyoxyéthylème: copolymères dont les séquences ont des longueurs tres inégales," *Makromol. Chem.,* to be published.

8. Perret, R. and Skoulios, A., "Etude de la cristallisation des copolymères triséquencés polycaprolactone-polyoxyéthylène: copolymères dont les séquences ont des longueurs voisines," *Markromol. Chem.,* to be published.

9. Skoulios, A., Helffer, P., Gallot, Y., and Selb, J., "Solubilization and Chain Conformation in a Block Copolymer System," *Makromol. Chem.,* 148 (1971), 305.

10. Guinier, A., *X-Ray Crystallographic Technology,* London: Hilger & Watts (1952).

11. Terrisse, J., "Contribution à l'étude des structures mésomorphes monocristallines de copolymères à séquences amorphes," thèse, Université Louis Pasteur de Strasbourg, 1973.

12. Grosius, P., Gallot, Y., and Skoulios, A., "Copolymères séquencés polystyrène/polyvinyl-2 pyridine: synthèse caractérisation et étude des phases mésomorphes," *Makromol. Chem.,* 127 (1969), 94.

13. Douy, A., Mayer, R., Rossi, J., and Gallot, B., "Structure of Liquid Crystalline Phases from Amorphous Block Copolymers," *Mol. Cryst. Liquid Cryst.,* 7 (1969), 103.

14. Grosius, P., Gallot, Y., and Skoulios, A., "Influence de la température sur la structure et les paramètres géometriques des copolymères séquencés polystyrène-polyvinyl-pyridine," *C.R. Acad. Sci., Ser. 3,* 270 (1970), 1381.

SESSION IV

PHYSICAL PROPERTIES

MODERATOR: TURNER ALFREY, JR.
Dow Chemical Company
Midland, Michigan

8. Microphase Separation and Block Size

SONJA KRAUSE

Rensselaer Polytechnic Institute
Troy, New York

ABSTRACT

The conditions necessary for microphase separation in block copolymers and in mixtures of block copolymers with one of the corresponding homopolymers are reviewed. These conditions depend on the composition of the block copolymer, its molecular weight, the number of blocks per molecule, the interaction parameter between the corresponding homopolymers, the molecular weight of the added homopolymer, and, when one component is crystallizable, its enthalpy and entropy of crystallization. The calculated results are compared with experimental data. Experimentally determined glass-transition temperatures in block copolymers are discussed briefly in terms of polymer compatibility and the small size of many microphases.

Most workers who have been interested in devising theoretical treatments of microphase separation in block copolymers have been drawn, for the most part, toward details of morphology, such as the exact dimensions of the microphases formed by block copolymers of different composition, molecular weight, and total number of blocks. Some workers have investigated the kinds of block copolymers that exhibit the various equilibrium types of microphases found experimentally: spherical inclusions of one block in a matrix of the other; cylindrical inclusions of one block in a matrix of the other; and lamellar or layered structures. These workers have included Meier [1], Bianchi et al. [2], Leary and Williams [3], Marker [4], LeGrand [5], Inoue et al. [6,7], and Pouchly et al. [8].

The approach used by many of the workers cited above has been microscopic; i.e., calculations on the morphology of single molecules form the basis of their theories. This approach is necessary when microphase dimensions are of interest [1–3], although Pouchly et al. [8] used an excluded volume approach in their treatment of isolated A–B block copolymer molecules.

My own approach to the prediction of microphase separation has been macro-

143

scopic, using a strictly thermodynamic approach. The main advantage of this approach is the fact that it predicts microphase separation independently of morphology, and that a single set of assumptions allows predictions to be made for crystallizable and noncrystallizable microphases, for diblock, triblock, or n-block copolymers, and for mixtures of block copolymers with one of the corresponding homopolymers. This independence of morphology is also the main disadvantage of this treatment, since it allows neither a prediction of morphology nor a calculation of microphase dimensions. Nevertheless, the predictions made by this treatment are in agreement with available experimental data, as shown below.

There would not be a conference on block copolymers approximately once every year if these materials were of purely theoretical interest. The existence of covalently bonded phases in these samples gives them unique properties leading to such products as the thermoplastic elastomers exemplified by Shell's Kraton series of polymers. The Kraton polymers are triblock copolymers of styrene and butadiene, with a butadiene block in the center of each molecule and a styrene block on each end. In a solid sample of this material, the styrene blocks from many molecules apparently combine to form spherical or cylindrical microphases in a polybutadiene matrix made up of the center block from each molecule. The polystyrene microphases have high glass-transition temperatures and are in the glassy state at room temperature. The polybutadiene matrix, on the other hand, has a low glass-transition temperature and is therefore rubbery at room temperature. The polystyrene microphases act as cross-links for the rubber at room temperature, but, unlike true chemical cross-links, they soften at higher temperatures, thus allowing the plastic to be formed and molded at the higher temperature. An ordinary cross-linked rubber, on the other hand, cannot be formed or molded after cross-linking is complete. An understanding of the circumstances under which microphase separation in such block copolymers takes place should be helpful in the search for other thermoplastic rubbers and other materials formed from block copolymers.

Furthermore, since the presence of two phases with two different glass-transition temperatures is of importance in block copolymers of commercial interest, an investigation of the effect of block copolymer size, compatibility, and microphase dimensions on the values of the observed glass-transition temperatures appears necessary. Some literature data that illuminate these questions are discussed below.

Assumptions

The block copolymers considered here are monodisperse samples in which every molecule has the same molecular weight, the same average composition in terms of the number of blocks of each type (A and B), and the same sequence distribution in terms of the number of blocks of each type (A and B). The blocks are fairly long, so that one may assume that there is no change in molecular disorder within each block *except* directly at the very sharp boundary between microphases unless one of the blocks (always block A in this paper) is crystallizable. Both the change in molecular disorder and the volume change on crystallization are encompassed in the entropy of crystallization. Complete separation between the microphases is assumed under all circumstances, whether or not the A blocks crystallize, and whether or not homopolymer A is present.

The homopolymer HA, considered here, is also monodisperse in molecular weight and dissolves only in microphase A after the microphase separation occurs. This assumption appears reasonable if the work of Inoue [5,7], Molau [12], and McIntyre [13] is considered; electron micrographs of homopolymer-block copolymer mixtures show complete microphase separation with sharp boundaries between microphases just like electron micrographs of the block copolymer alone. However, there is evidence that some mixing of different repeat units occurs even in some pure block copolymers [14], and there is no a priori reason to assume that homopolymer A will never mix with microphase B after microphase separation occurs. A more complete treatment of microphase separation requires consideration of diffuse boundaries between microphases with a concomitant minimization of the free energy in order to determine the amount of mixing; such a treatment has been reported by Leary and Williams [3] for pure A-B-A block copolymers. The simpler approximation, involving sharp boundaries between microphases and no mixing of any A and B units after microphase separation, has nevertheless been used in this work, and it is expected to yield qualitative trends correctly. The effect of added homopolymer, HA, is considered only in the case of noncrystallizable systems.

Enthalpy

If these assumptions and the lattice theory of solutions are used, the enthalpy change on microphase separation when no crystallization occurs is

$$\Delta H = - \frac{kTV}{V_r} v_A^M v_B^M \chi_{AB} \left(1 - \frac{2}{z}\right), \tag{1a}$$

where V is the total volume of the system, T is the absolute temperature, k is the Boltzmann constant, V_r is the volume of a lattice site, z is the coordination number of the lattice, v_A^M and v_B^M are volume fractions of monomer repeat units A and B in the total mixture, respectively, and χ_{AB} is the interaction parameter between A units and B units. Equation (1a) is the Hildebrand [15]-Scatchard [16]-Van Laar [17] heat of mixing that is lost on microphase separation if the fixed composition of each polymer chain is considered. For each monomer unit, except end units, in each polymer chain, only $z - 2$ contacts are random before microphase separation, since two contacts are fixed by the composition of the chain. It is also assumed that the only contacts between A and B after microphase separation are at the surfaces between microphases and consist only of A–B junctions within block copolymer molecules. This assumption restricts the discussion to small and moderate volume fractions of added homopolymer.

In the absence of added homopolymer, and when the microphases of block A are crystalline, the enthalpy change on microphase separation is

$$\Delta H = - \frac{kTN_c}{V_r} V_A \, n_A^c \, v_B^c \, \chi_{AB} \left(1 - \frac{2}{z}\right) + n_A^c N_c \frac{\Delta H_{\text{cryst}}}{N_o} , \qquad (1b)$$

where V_r is still the volume of a lattice site in the amorphous system, V_A is the amorphous volume of a repeat unit of A, n_A^c is the number of A units in each copolymer molecule, v_B^c is the volume fraction of B in the copolymer, N_c is the total number of copolymer molecules in the system, N_o is Avogadro's number and ΔH_{cryst} is the measured molar enthalpy of crystallization per unit of A for homopolymers of A.

The first term on the right side of Equation (1b) is the same as Equation (1a) would be in the absence of added homopolymer; the second term on the right side of Equation (1b) is the additional enthalpy change when the A-microphases crystallize.

Entropy

The total entropy change when microphase separation occurs in the absence of crystallization is

$$\frac{\Delta S}{k} = N_c \ln \left(v_A^M\right)^{v_A^c} \left(v_B^M\right)^{v_B^c} + N_{HA} \ln v_A^M$$

$$- 2N_c(m - 1) \left(\frac{\Delta S_{\text{dis}}}{R}\right) + N_c \ln(m - 1) , \qquad (2a)$$

where N_{HA} is the number of homopolymer molecules in the system, v_A^c is the volume fraction of monomer A in the copolymer molecules, and m is the number of blocks in each block copolymer molecule. The first term on the right side of Equation (2a) is concerned with the decrease in available volume for each block in the copolymer molecules after microphase separation, assuming that there is no volume change on mixing. The second term on the right side of Equation (2a) is concerned with the decrease in available volume for the homopolymer molecules after microphase separation. Terms of this type were discussed in a previous paper [9]. The third term on the right side of Equation (2a) is the additional entropy decrease caused by immobilization at the microphase surface of the A and B units that link the blocks in each copolymer molecule. The fourth term arises for block copolymers containing three or more blocks because of the large number of sites available for the block-linking segments on the surface between microphases. These last two terms have also been discussed previously [10], especially the values that could be used for $(\Delta S_{dis}/R)$, the disorientation entropy gain on fusion per segment of a polymer. In agreement with previous calculations, a value of $(\Delta S_{dis}/R) = 1.0$ is used in this paper; actual values of $(\Delta S_{dis}/R)$ range from 0.85 to 4.3 for various polymers [18].

In the absence of added homopolymer, and when the microphases of block A are crystalline, the entropy change on microphase separation is

$$\frac{\Delta S}{k} = N_c \ln (v_a^c)^{v_A^c} (v_B^c)^{v_B^c} - 2N_c (m-1) \left(\frac{\Delta S_{dis}}{R}\right)$$

$$+ N_c \ln (m-1) + N_c n_A^c \left(\frac{\Delta S_{cryst}}{R}\right) , \qquad (2b)$$

where ΔS_{cryst} is the measured molar entropy of crystallization per unit of A for homopolymers of A.

The first three terms on the right side of Equation (2b) are the same as Equation (2a) would be in the absence of added homopolymer; the last term on the right side of Equation (2b) is the additional entropy change when the A-microphases crystallize.

Free Energy

The free energy change on microphase separation for the system of N_c copolymer molecules and N_{HA} homopolymer molecules occupying volume V is $\Delta G = \Delta H - T\Delta S$, where ΔH and ΔS are the expressions in Equations (1a) and (2a), respectively. When $\Delta G = 0$, the separated system is in equilibrium with the completely mixed system; solution of this equation allows calculation of critical values of χ_{AB}, $(\chi_{AB})_{cr}$, for various ratios of N_{HA} to N_c, for various values v_A^c and m, and for various degrees of polymerization of the two polymers. Since

$$V = N_c (V_A n_A^c + V_B n_B^c) + N_{HA} V_A n_{HA} \qquad (3)$$

and

$$v_A^M = \frac{N_{HA} V_A n_{HA} + N_c V_A n_A^c}{V} , \qquad (4)$$

where V_A and V_B are the volumes of A and B repeat units, respectively, n_A^c and n_B^c are the number of A and B units in each copolymer molecular, respectively, and n_{HA} is the number of A units in each homopolymer molecule; we obtain

$$(\chi_{AB})_{cr} = \frac{zV_r}{(z-2)V_B v_A^M n_B^c} \left[-\ln(v_A^M)^{v_A^c} (v_B^M)^{v_B^c} - \frac{N_{HA}}{N_c} \ln v_A^M \right.$$

$$\left. + 2(m-1)\left(\frac{\Delta S_{dis}}{R}\right) - \ln(m-1) \right]. \tag{5a}$$

When Equation (5a) is used to make predictions, it is convenient to let $V_r = V_A = V_B$ [10], and to set a convenient value for z, possibly $z = 8$, and for $(\Delta S_{dis}/R)$, possibly the value 1.0 mentioned above.

Equation (5a) was derived using essentially an approach suitable for a one-component system, although a mixture of homopolymer with block copolymer is a two-component system. This approach was possible because it was assumed that the homopolymer HA did not mix with microphase B after microphase separation; i.e., the system was not allowed to act like an ordinary two-component system.

In the absence of added homopolymer, and when the microphases of A are crystalline, setting $\Delta G = 0$ also gives a value of $(\chi_{AB})_{cr}$:

$$(\chi_{AB})_{cr} = \frac{zV_r}{(z-2)V_A n_A^c v_B^c} \left[n_A^c \left\{ \frac{\Delta H_{cryst}}{RT} - \frac{\Delta S_{cryst}}{R} \right\} \right.$$

$$\left. -\ln(v_A^c)^{v_A^c} (v_B^c)^{v_B^c} + 2(m-1)\left(\frac{\Delta S_{dis}}{R}\right) - \ln(m-1) \right]. \tag{5b}$$

When using Equation (5b), it is also useful to let $V_r = V_A$ and to set $z = 8$ and $(\Delta S_{dis}/R) = 1.0$. The entropy and enthalpy of crystallization used in Equation (5b) should be experimentally determined values for the blocks of interest, preferably at the temperature of interest. Since such values are often available only at the equilibrium melting point of the homopolymer, those values would usually have to be used.

Predictions and Comparison with Experiment

Noncrystallizable Monodisperse Block Copolymers

Predictions of $(\chi_{AB})_{cr}$ for noncrystallizable monodisperse block copolymers with degree of polymerization 400 and $v_A^c = 0.25$ and 0.50 comprised of 2 to 10 total blocks were made in a previous paper [10]. Using $z = 8$, $(\Delta S_{dis}/R) = 1.0$, and Equation (5a) with no added homopolymer, we can predict the values of $(\chi_{AB})_{cr}$ shown in Table I. Since $(\chi_{AB})_{cr}$ is the lowest value of the interaction parameter between the corresponding homopolymers for which a particular block copolymer will exhibit phase separation, a low value of $(\chi_{AB})_{cr}$ indicates

TABLE I

$(\chi_{AB})_{cr}$ for Noncrystallizable Monodisperse Block Copolymers

Total Degree of Polymerization	m	$(\chi_{AB})_{cr}$ if $v_A^c = 0.25$	$(\chi_{AB})_{cr}$ if $v_A^c = 0.50$
400	2	0.047	0.036
	3	0.071	0.053
	4	0.100	0.075
	6	0.161	0.121
800	2	0.023	0.018
	3	0.034	0.027
	4	0.048	0.037

a great likelihood of microphase separation for an arbitrarily chosen block copolymer of the proper degree of polymerization, composition, and number of blocks. Therefore, Table I shows that a block copolymer with a higher total degree of polymerization is more likely to show microphase separation and, for any particular total degree of polymerization, a diblock copolymer is more likely to exhibit microphase separation than a triblock copolymer. The triblock copolymer is more likely to exhibit microphase separation than a tetrablock copolymer, and so on; and a 50/50 copolymer is more likely to separate into microphases than a 25/75 copolymer.

It is interesting to compare the calculations from Equation (5a) with at least one set of experimental data; and the data on block copolymers of styrene and α-methyl styrene obtained by Baer [19] have been selected for discussion here. Baer prepared triblock copolymers whose number average molecular weights were all above 10^5; i.e., the total degree of polymerization of each sample was greater than 1,000. The composition of the samples was approximately 50/50 ($v_A^c = 0.5$). The molded unfractionated block copolymers were transparent and showed only a single glass-transition temperature intermediate between those of the corresponding homopolymers. These data indicate that there was no microphase separation in these samples. When samples of the two homopolymers were blended (in solution), however, cloudy samples with two damping maxima were obtained. These data indicate that the homopolymers were incompatible and did not blend into a single phase on mixing.

In order to use Equation (5a) on this system, it is necessary to first calculate χ_{AB}. This was done using Hoy's [20] version of Small's [21] tables and a density of 1.049 for polystyrene and 1.066 for α-methyl styrene [22] to calculate, first, Hildebrand solubility parameters [23] for the polymers and, then, χ_{AB} at 25°C. The calculated value was 0.002, much smaller than the $(\chi_{AB})_{cr}$ for total degree of polymerization 800, $m = 3$, $v_A^c = 0.5$ shown in Table I, namely, 0.027. Even for total degree of polymerization 1,200, the value of $(\chi_{AB})_{cr}$ calculated from Equation (5a) is as high as 0.018. For a triblock copolymer with 50/50 composition, $(\chi_{AB})_{cr}$ does not drop to 0.002 until the total degree of polymerization is 10,000, corresponding to a molecular weight of the order of 10^6. Further work with this system would provide a good test of this theoretical treatment.

It can be shown that Baer's [19] results with a mixture of the corresponding homopolymers are also reasonable, even though he gave no information on the molecular weights of his homopolymers. The Flory–Huggins [24] theory can be used to calculate the limit of molecular weight at which a mixture of homopolymers with $\chi_{AB} = 0.002$ is miscible in all proportions. This occurs at degree of polymerization 1,000, molecular weights of the order of 10^5. If the molecular weights of Baer's homopolymers were greater than 10^5, phase separation was to be expected.

Mixtures of Block Copolymers with One of the Corresponding Homopolymers

Tables of predicted values of $(\chi_{AB})_{cr}$ from Equation (5b) were given in a previous paper [11]. Without repetition of the tabulated values, the results of those calculations are reviewed here. The addition of homopolymer A to a block copolymer of A and B lowers $(\chi_{AB})_{cr}$, i.e., increases the likelihood of microphase separation, in practically all cases. There are a few cases in which the addition of a homopolymer to the block copolymer increases $(\chi_{AB})_{cr}$ very slightly, i.e., makes the system slightly more compatible; in the case of diblock copolymers, this happens only when the degree of polymerization of the homopolymer is equal to or less than one-quarter the degree of polymerization of the block copolymer; in the case of triblock or tetrablock copolymers, this happens only when the degree of polymerization of the homopolymer is equal to or less than one-eighth the degree of polymerization of the block copolymer. Addition of high-molecular-weight homopolymer, on the other hand, can decrease $(\chi_{AB})_{cr}$ appreciably, i.e., greatly increase the likelihood of microphase separation.

There are no published data, as far as I know, that can be compared with these predictions. The calculations were made because block copolymer samples, as synthesized, often contain one or more of the corresponding homopolymers. It is therefore of interest to investigate the effects of the presence of such homopolymers on the phase separation of the block copolymer.

It must be emphasized here that the present theoretical treatment deals only with the conditions under which a homopolymer–block copolymer mixture changes from a true solution to a system in which the block copolymer has undergone microphase separation with all of the homopolymer dissolved in the microphase made up of the corresponding blocks. The present work does not deal with the situation studied experimentally by Inoue [7], in which the conditions under which added homopolymer forms a microphase of its own instead of entering the microphases of the block copolymer are of interest. The emulsifying effect of block copolymer on the corresponding homopolymers (also studied by Molau [12]) is not covered by this theoretical treatment.

Block Copolymers with One Crystallizable Block

At temperatures far below the melting point of high-molecular-weight samples of the corresponding homopolymer, $(\chi_{AB})_{cr}$ as calculated from Equation (5b) depends mainly on the enthalpy and entropy of crystallization of the polymer. The term

$$n_A^c \left[(\Delta H_{cryst}/RT) - (\Delta S_{cryst}/R) \right]$$

is then a highly negative term and dominates Equation (5b). This can be illustrated using ethylene oxide as A and calculating $(\chi_{AB})_{cr}$ for a 50/50 diblock copolymer with $(\Delta S_{dis}/R) = 1.0$, $(\Delta S_{cryst}/R) = 2.9$ [18], $\Delta H_{cryst} = -2,000$ calories per repeat unit [18], and a temperature of 25°C. In that case,

$$(\chi_{AB})_{cr} = \frac{7.18}{n_A^c} - 1.33 \quad . \tag{6}$$

When n_A^c, the number of units of A in the ethylene oxide block, exceeds six, $(\chi_{AB})_{cr}$ becomes negative; i.e., microphase separation is predicted no matter what the chemical nature of block B. About ten years ago, Skoulios and his coworkers [25,26] found that very low-molecular-weight copolymers of ethylene oxide with propylene oxide [25] and with styrene [26] exhibited microphase separation with crystalline regions of ethylene oxide. The total molecular weight of some of the block copolymers studied was only 4,000.

Glass Transitions in Block Copolymers

It is known [27,28] that polymer mixtures exhibit two glass-transition temperatures when the polymers are incompatible, and only a single glass-transition temperature when the polymers are compatible. Furthermore, Krause and Roman [29] found that a mixture of compatible polymers, poly(isopropyl acrylate) and poly(isopropyl methacrylate), has the same glass-transition temperature as a random copolymer of the same composition. On occasion, it is found that a mixture of two random copolymer samples made up from the same two monomers but differing in composition shows two glass-transition temperatures [30], as would be expected if phase separation occurs because of the incompatibility of the two samples.

Many glass-transition-temperature data have been obtained using samples of block and graft copolymers. Almost always, such samples have two glass-transition temperatures [31] because of the microphase separation within the samples. The case of triblock copolymers of styrene and α-methyl styrene, which had only a single glass-transition temperature intermediate between those of the corresponding homopolymers and were therefore assumed to have no microphase separation [19], has already been discussed.

A number of authors have noticed that the glass-transition temperature of a particular microphase depends on the length of the blocks making up that microphase. For styrene–butadiene block copolymers, Kraus et al. [31] noticed that the temperature of the loss peak attributed to the styrene blocks decreased from 104°C to 60°C when the styrene block length decreased by a factor of about 4 (exact molecular weights of the samples were unknown). Childers and Kraus [32] felt that the lowering of the glass-transition temperature of the styrene phase with decreasing block length might possibly be attributed to some mixing of butadiene segments with those of styrene, i.e., increased compatibility of the blocks, as the styrene block length decreased. The first explanation is reasonable since the variation of glass-transition temperature with molecular weight is well known in homopolymers, at least up to a number average molecular weight of about 70,000 [33–36]. The second explanation is equally reasonable, even though the theoretical treatment of microphase separation advanced here does not allow for partial mixing in the microphases. Table I shows that $(\chi_{AB})_{cr}$ for a triblock copolymer with total degree of polymerization 400 and $v_A^c = 0.25$ is of the order of 0.07. If all assumptions are reasonable, and if $\chi_{AB} = 0.07$ is used for the styrene–butadiene system [11], a low-molecular-weight triblock copolymer with 25/75 composition would not be expected to show microphase separation at all. It is therefore not unreasonable to suppose that partial mixing of the blocks can exist in this system, especially at low molecular weights.

Bares and Pegoraro [37] analyzed the observed glass-transition temperatures in ethylene–propylene copolymers with vinyl chloride grafts in terms of mixing at the surfaces of the vinyl chloride microphases due to surface contacts with ethylene and propylene monomers with no mixing in the microphase interiors. This explanation gives noticeable variations in glass transition temperatures only if the microphases are extremely small; Bares [37] calculated the diameters of the microphases as 25 monomer units if spherical and 10 monomer units if cylindrical.

Further work on the variation of glass-transition temperatures of the microphases in block copolymer samples with compatibility, block size, and microphase dimensions seems necessary. More than one measuring technique will have to be used, since evidence [38–40] exists that the sensitivity of different measuring techniques (dynamic mechanical measurements, differential calorimetry, etc.) toward the glass transitions in mixtures of homopolymers depends on the physical dimensions of the phases.

Acknowledgment

The work reported in this paper was supported in part by the National Aeronautics and Space Administration under Grant NGL-33-018-003, and in part by the National Science Foundation under Grant GH-32955.

References

1. Meier, D.J., "Theory of Block Copolymers. I. Domain Formation in A–B Block Copolymers," *J. Polym. Sci., Part C,* 26 (1969), 81.
2. Bianchi, V., Pedemonte, E., and Turturro, A., "Statistical Thermodynamics of Styrene–Butadiene Block Copolymers," *J. Polym. Sci., Part B,* 7 (1969), 785.
3. Leary, D.F. and Williams, M.C., "Statistical Thermodynamics of A–B–A Block Copolymers: I," *J. Polym. Sci., Part B,* 8 (1970), 335.
4. Marker, L., "Phase Equilibria and Transition Behavior of Block Polymers–A Simple Model," *Polym. Prepr., Amer. Chem. Soc., Div. Polym. Chem.,* 10, no. 2 (1969), 524.
5. LeGrand, D.G., "The Effect of Interactions on the Morphology and Properties of Block Copolymers," *Polym. Prepr., Amer. Chem. Soc., Div. Polym. Chem.,* 11, no. 2 (1970), 434.
6. Inoue, T., Soen, T., Hashimoto, T., and Kawai, H., "Thermodynamic Interpretation of Domain Structure in Solvent-Cast Films of A–B Type Block Copolymers of Styrene and Isoprene," *J. Polym. Sci., Part A-2,* 7 (1969), 1283.
7. Inoue, T., Soen, T., Hashimoto, T., and Kawai, H., "Studies on Domain Formation of the A–B Type Block Copolymer from its Solutions. Ternary Polymer Blend of the Styrene–Isoprene Block Copolymer with Polystyrene and Polyisoprene," *Macromolecules,* 3 (1970), 87.
8. Pouchly, J., Zivney, A., and Sikora, A., "Incompatibility of Blocks in an Isolated Molecule of A–B Block Copolymers," *J. Polym. Sci., Part A-2,* 10 (1972), 151.
9. Krause, S., "Microphase Separation in Block Copolymers: Zeroth Approximation," *J. Polym. Sci., Part A-2,* 7 (1969), 249.
10. Krause, S., "Microphase Separation in Block Copolymers. Zeroth Approximation Including Surface Free Energies," *Macromolecules,* 3 (1970), 84.
11. Krause, S., "Microphase Separation in Mixtures of Block Copolymers with the Corresponding Homopolymers: Zeroth Approximation," in *Colloidal and Morphological Behavior of Block and Graft Copolymers,* G.E. Molau, ed., New York: Plenum Press (1971), 223.
12. Molau, G.E. and Wittbrodt, W.M., "Colloidal Properties of Styrene–Butadiene Block Copolymers," *Macromolecules,* 1 (1968), 260.
13. McIntyre, D. and Campos-Lopez, E., "The Macrolattice of a Triblock Polymer," *Macromolecules,* 3 (1970), 322.
14. Beecher, J.F., Marker, L., Bradford, R.D., and Aggarwal, S.L., "Morphology and Mechanical Behavior of Block Polymers," *Polym. Prepr., Amer. Chem. Soc., Div. Polym. Chem.,* 8, no. 2 (1967), 1532.
15. Hildebrand, J.H., "Solubility. XIV. Experimental Tests of a General Equation for Solubility," *J. Amer. Chem. Soc.,* 57 (1935), 866.
16. Scatchard, G., "Equilibria in Non-Electrolyte Solutions in Relation to the Vapor Pressure and Densities of the Components," *Chem. Rev.,* 8 (1931), 321.

17. Van Laar, J.J., "Relation Between the Deviation of the Vapor Pressure of Binary Mixtures of Normal Compounds from the Straight Line and the Heat of Mixing in the Liquid State," *Z. Phys. Chem., Abt. A.*, 137 (1928), 421.
18. Bondi, A., *Physical Properties of Molecular Crystals, Liquids and Glasses*, New York: Wiley (1968).
19. Baer, M., "Anionic Block Polymerization. II. Preparation and Properties of Block Copolymers," *J. Polym. Sci., Part A*, 2 (1964), 417.
20. Hoy, K.L., "New Values of the Solubility Parameters from Vapor Pressure Data," *J. Paint Technol.*, 42 (1970), 76.
21. Small, P.A., "Some Factors Affecting the Solubility of Polymers," *J. Appl. Chem.*, 3 (1953), 71.
22. Lewis, O.G., *Physical Constants of Linear Polymers*, New York: Springer-Verlag (1968).
23. Hildebrand, J.H. and Scott, R.L., *Solubility of Nonelectrolytes*, 2d ed., New York: Reinhold (1950), and 3d ed., paper, New York: Dover (1964).
24. Flory, P.J., *Principles of Polymer Chemistry*, Ithaca, N.Y.: Cornell University Press (1953).
25. Tsouladze, G. and Skoulios, A., "La Structure des colloides D'Association. IX. Caractère amphipathique et phases mésomorphes des copolymères oxyde de propylène-oxide d'éthylène," *J. Chim. Phys.*, 60 (1963), 626.
26. Skoulios, A. and Finaz, G., "La Structure des colloides D'Association. VII. Caractère amphipathique et phases mésomorphes des copolymères séquencés styrolène-oxyde d'éthylène," *J. Chim. Phys.*, 59 (1962), 473.
27. Buchdahl, R. and Nielsen, L.E., "Multiple Dispersion Regions in Rigid Polymeric Systems," *J. Polym. Sci.*, 15 (1955), 1.
28. Bartenev, G.M. and Kongarov, G.S., "Dilatometric Determination of Compatibility of Polymers," *Vysokomol. Soedin.*, 2 (1960), 1692. (In Russian)
29. Krause, S. and Roman, N., "Glass Temperatures of Mixtures of Compatible Polymers," *J. Polym. Sci., Part A*, 3 (1965), 1631.
30. Chandler, L.A. and Collins, E.A., "Multiple Glass Transitions in Butadiene–Acrylonitrile Copolymers," *Polym. Prepr., Amer. Chem. Soc., Div. Polym. Chem.*, 9, no. 2 (1968), 1416.
31. Kraus, G., Childers, C.W., and Gruner, J.T., "Properties of Random and Block Copolymers of Butadiene and Styrene. I. Dynamic Properties and Glass Transition Temperatures," *J. Appl. Polym. Sci.*, 11 (1967), 1581.
32. Childers, C.W. and Kraus, G., "Properties of Random and Block Copolymers of Butadiene and Styrene. III. Three Sequence Styrene–Butadiene–Styrene Block Polymers," *Rubber Chem. Technol.*, 40 (1967), 1183.
33. Fox, T.G. and Flory, P.J., "Second-Order Transition Temperatures and Related Properties of Polystyrene. I. Influence of Molecular Weight," *J. Appl. Phys.*, 21 (1950), 581.
34. Ueberreiter, K. and Kanig, G., "Self-Plasticization of Polymers," *J. Colloid Sci.*, 7 (1952), 569.
35. Beevers, R.B. and White, E.F., "Physical Properties of Vinyl Polymers. Part 1—Dependence of The Glass-Transition Temperature of Polymethylmethacrylate on Molecular Weight," *Trans. Faraday Soc.*, 56 (1960), 744.
36. Beevers, R.B. and White, E.F., "Physical Properties of Vinyl Polymers. Part 2—The Glass Transition Temperature of Block and Random Acrylonitrile and Methyl Methacrylate Copolymers," *Trans. Faraday Soc.*, 56 (1960), 1529.
37. Bares, J. and Pegoraro, M., "Properties of Ethylene–Propylene–Vinyl Chloride Graft Copolymers. II. Viscoelasticity and Composition of Graft Copolymer and Composite," *J. Polym. Sci., Part A-2*, 9 (1971), 1287.

38. Bank, M., Leffingwell, J., and Thies, C., "The Influence of Solvent Upon the Compatability of Polystyrene and Poly(vinyl methyl ether), *Macromolecules,* 4 (1971), 43.
39. Stoelting, J., Karasz, F.E., and MacKnight, W.J., "Dynamic Mechanical Properties of Poly(2,6-Dimethyl-1,4 Phenylene Ether)–Polystyrene Blends," *Polym. Prepr., Amer. Chem. Soc., Div. Polym. Chem.,* 10, no. 2 (1969), 628.
40. Sperling, L.H., Taylor, D.W., Kirkpatrick, M.L., George, H.F., and Bardman, D.R., "Glass-Rubber Transition Behavior and Compatibility of Polymer Pairs: Poly(ethyl Acrylate) and Poly(methyl Methacrylate). *J. Appl. Polym. Sci.,* 14 (1970), 73.

9. Block Polymers as Microcomposites

S. L. AGGARWAL, R. A. LIVIGNI, LEON F. MARKER,
AND T. J. DUDEK

The General Tire and Rubber Company
Akron, Ohio

ABSTRACT

The phase-separated systems produced from block and graft polymers can be viewed as a novel form of microcomposites. These systems are unique, however, in that their morphology can be conveniently altered through techniques used in their synthesis and the manner in which they are subsequently treated. We have shown that anionic copolymerization techniques can be used to prepare copolymers of specified composition and controlled sequence length. These polymers are particularly sensitive to morphological variation. A thermodynamic argument has been developed to explain many of the morphological features and transition behavior of these microcomposites. An attempt is made to demonstrate the applicability of composite theory in describing their mechanical behavior.

In his book on copolymerization, Alfrey [1] pointed out that the presence of long sequences of a particular monomer in a copolymer may result in incompatibility at a submicroscopic level. He predicted that such materials may show properties quite different from either a random copolymer or a macroscopic blend of two homopolymers. In retrospect, from what is now known about block polymers, this prediction, made almost twenty years ago, seems prophetic.

A number of studies, including some of the early work from this laboratory [2], have shown that the block polymers in the solid state have a multiphase structure. The unique features of this structure are that the dispersed phase is of microscopic or submicroscopic dimensions and that the phases are interconnected by definite chemical bonds. It is in this context that we wish to consider block polymers as microcomposites to distinguish them from conventional composite materials.

For the first time in polymeric materials, block polymers offer a means of obtaining microcomposites with some control of the morphology and orientation

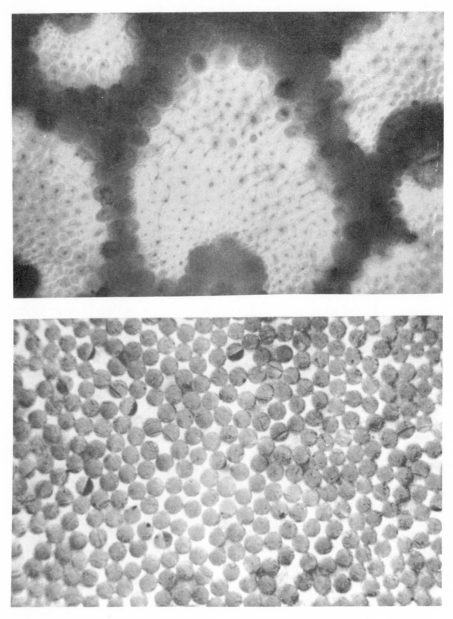

Figure 1. Natural and synthetic composite materials: micrographs of bamboo-reinforced and filament-wound glass-reinforced plastics.

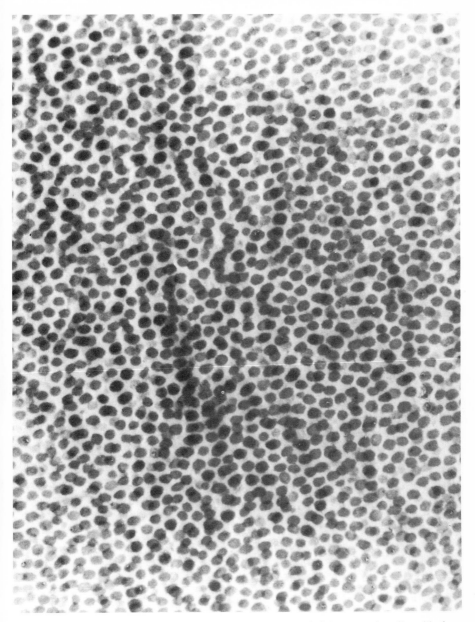

Figure 2. Electron micrograph of an ultrathin section of a 90/10 styrene–butadiene block polymer.

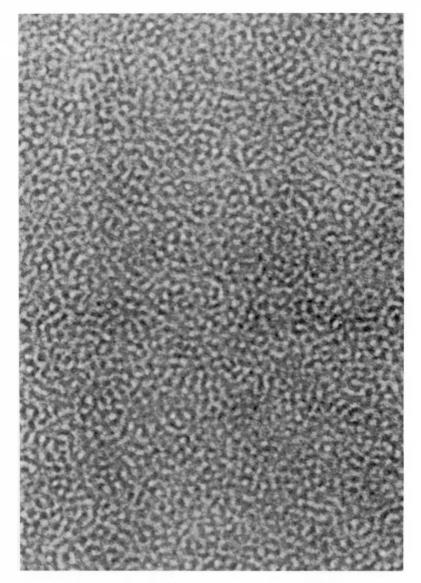

Figure 3. Electron micrograph of Kraton 101 cast from a 90/10 THF–MEK solution.

of the phases. The study of block polymers thus offers exciting possibilities for obtaining strong polymer materials, because most of the strong materials, both natural and synthetic, are composite materials with two-phase or multiphase structure.

The two-phase structure in two of the strongest materials for their weight, bamboo- and glass-fiber-reinforced plastics, is shown in Figure 1. In bamboo, highly oriented, high-strength cellulose fibers (dark spots in the micrographs) are embedded in lignin (gray areas), which forms the continuous phase. In the fiber-reinforced plastics, the glass fibers form the discontinuous phase, and a plastic resin the continuous phase.

In many cases, block polymers separate into highly organized, regular geometrical arrangements. Under some circumstances, well-separated rubber domains can be obtained that appear to be of constant size and shape. An example of

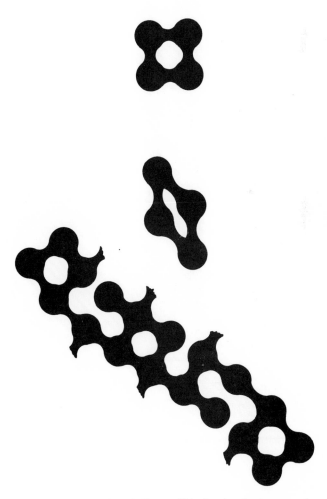

Figure 4. Schematic of morphology in Kraton 101, showing short-range order and particle interconnection.

Figure 5. Electron micrograph of an ultrathin section cut parallel to extruded and annealed plug (from [4]).

this is the micrograph of an ultrathin section of a 90/10 styrene–butadiene–styrene block polymer shown in Figure 2. In this and subsequent micrographs, the osmium-tetroxide-stained polydiene phase is dark; the polystyrene phase is light.

Early work on Shell's Kraton [2] thermoplastic rubber suggested that the polystyrene domains were spherical. In Figure 3, one of our early micrographs [2], there is an indication that the polystyrene domains are interconnected and that there are local regions ordered in a simple cubic arrangement. Such ordered structures are shown schematically in Figure 4. The small-angle X-ray-scattering work of McIntyre and Campos-Lopez [3] indicated that a good deal of long-range order exists in this system and that domains of regular size arrange themselves into a superlattice structure. Keller and his associates in a number of recent studies [4] have shown that there is a preference in this system for rodlike structures and that these parallel rods are arranged in a hexagonal array. Figure 5 is one of the micrographs from their studies demonstrating rod formation from extruded Kraton samples. LaFlair [5] has studied the morphology of similar polymers and has demonstrated that rodlike domain structures have been mistakenly interpreted as spheres when viewed in cross-section. All of these studies suggest that under suitable conditions composite structures with fairly regular geometry of the dispersed phase can be realized in block polymers.

In this paper, the following topics are discussed:

1. The application of recently developed theories for composite materials to block polymers having morphology of regular geometry.

2. The effect of solvents on anionic copolymerization that results in block copolymers of varied but controlled structure and potentially provides a means of preparing microcomposites of varied morphologies.

3. The effect of the structure of block copolymers and the solvents from which films are prepared on the phase morphology and transition behavior of block copolymers.

4. Morphology, transition behavior, and stress–strain behavior of films of Kraton cast from different solvents.

5. The thermodynamics of phase separation in block polymers, which may be helpful in understanding the morphology and properties of these systems.

Applications of Composite Theories

Review of Theories of Micromechanics of Composites

A number of authors have demonstrated that the computation of properties of composite systems, such as elastic moduli, gas permeability, and thermal conductivity, can be subjected to rigourous analysis [6–8]. Solutions have been obtained using a number of geometries. Hashin, in a classical set of papers [6], made use of variational energy methods to establish upper and lower bounds for the relevant moduli for composite systems of irregular phase geometry, for systems containing randomly dispersed spheres, and for systems containing uniaxially oriented fibers. Hashin's bounds serve as a test for any more approximate theory. For a random assemblage of spheres, the system is isotropic, and only two material parameters are needed to characterize this system completely. He found that the upper and lower bounds for bulk and compression moduli were almost identical. For shear moduli, these bounds are close only when the moduli of matrix and particle are close.

The lower bound for the shear modulus is

$$\frac{G}{G_m} = 1 + \frac{15\,(1 - \nu_m)\left(\dfrac{G_p}{G_m} - 1\right) C}{7 - 5\nu_m + (8 - 10\nu_m)\left[\dfrac{G_p}{G_m} - \left(\dfrac{G_p}{G_m} - 1\right) C\right]}. \tag{1}$$

Here ν_m is the Poisson ratio of the matrix; G_p and G_m are shear moduli of particle and matrix, respectively; and C is the volume concentration of the spheres. This turns out to be identical to an expression derived by Kerner [9] by an entirely different method, which is a frequently used approximation for discontinuous composites.

Equation (1) is applied to calculate the shear modulus of Kraton 101, assuming that it is a microcomposite with spherical domains dispersed in a matrix of polybutadiene (a suggested morphology for films of Kraton 101 cast from benzene-heptane solvent). Kraton, as mentioned earlier, is the styrene-butadiene-styrene block polymer commercially available from Shell. It has a molecular weight of 78,000 and has 28 percent styrene by weight. The shear modulus of polystyrene [10] is assumed to be 1.13×10^{10} dynes/cm^2, and that of the rubber [11] 1×10^7 dynes/cm^2.

On substituting these values into Equation (1), we obtain a composite shear modulus of 1.85×10^7 dynes/cm^2. Experimental values are found to be much higher than this. For instance, for a sample of Kraton 101 cast from a benzene-heptane solution, a shear modulus of 10^8 dynes/cm^2 is obtained. The large discrepancy between theoretical and experimental values suggests that either the matrix modulus is much higher than the assumed value and/or that the domains deviate significantly from a spherical geometry. Both of these points are considered later.

Uniaxially oriented fiber-composite materials are transversely isotropic. The basic geometry of this system is shown in Figure 6. Five independent elastic constants would be needed to characterize completely the elastic behavior of such a material. The types of loading involved and the elastic constants associated with that loading are given in Figure 6.

Hashin [6] has shown that for four of these material constants, i.e., E_{11}, G_{12}, ν_{12}, and E_{22}, the upper and lower bounds are almost identical. Only for the shear modulus G_{23} are the upper and lower bounds widely divergent. Hill and

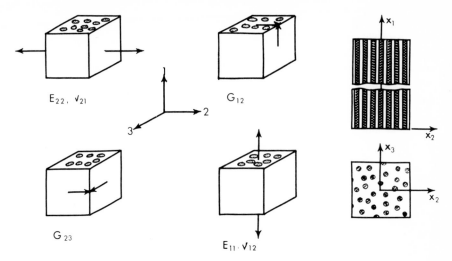

Figure 6. Types of loading and corresponding elastic constants (left); basic geometry of uniaxial composite and reference coordinate system (right).

Hermans [7,8] have used the self-consistent method, which is limited only in the accuracy of the model, to derive an alternative set of equations. Halpin and Tsai [12] have put these equations into the following convenient approximate form, which for E_{11}, G_{12}, ν_{12}, and E_{22} are equivalent to the Hashin's equations.

$$E_{11} = E_f V_f + E_m V_m \tag{2}$$

$$\nu_{12} = \nu_f V_f + \nu_m V_m \tag{3}$$

$$\frac{P}{P_m} = \frac{P_f/P_m + \zeta \left[1 + (P_f/P_m - 1)\nu_f\right]}{P_f/P_m - V_f(P_f/P_m - 1) + \zeta} \tag{4}$$

where

E_f, E_m = Young's modulus of fiber and matrix, respectively
ν_f, ν_m = Poisson ratio of fiber and matrix, respectively
V_f, V_m = volume fraction of fiber and matrix
$\quad\quad P = E_{22}, G_{12},$ or G_{23} for the composite
$\quad\quad P_m = E_m, G_m,$ or ν_m for the matrix
$\quad\quad P_f = E_f, G_f,$ or ν_f for the fiber, and
$\quad\quad \zeta$ = a parameter that accounts for the geometry of the filler and the mode of deformation

Equations (2) and (3) are the familiar "law of mixtures" for principal Young's modulus and principal extension ratio, respectively. Equation (4) is a condensed formula for determining the elastic constants E_{22}, G_{12}, or G_{23}. Use of Equations (2)–(4) is not restricted to uniaxial continuous fiber composites. Halpin and Tsai have shown the following:

1. Equation (4) can be applied to the calculation of the axial modulus of discontinuous composites by choosing an appropriate constant, ζ, which depends on the particle shape and aspect ratio.

2. The above equations can be extended to balanced ply laminates of discontinuous or continuous fiber composites by application of laminated plate theory [13] and by taking into account the angles between the ply orientation and the principal directions of the composite.

3. These concepts can be extended to randomly oriented fiber composites that are treated as quasi-isotropic materials [14]; i.e., the material is treated as a many-ply laminated structure having equal probability of all orientations.

Mechanical Properties of Rods and Sheets of Kraton with Oriented Polystyrene Domains

As mentioned earlier, Keller and his co-workers have studied the morphology and properties of rods of Kraton that had been extruded under special conditions and then annealed. They showed that the morphology of polystyrene in such samples was in the form of long polystyrene cylinders oriented in the extrusion direction and embedded in the matrix of polybutadiene [4]. We prepared extruded rods and oriented sheets from Kraton 101 and measured the moduli of these samples in tension, torsion, and flexure. The following is a brief description of the preparation of these samples, the experimental methods used for measurements of the various moduli, and the results of measurements on these samples.

1. *Preparation of samples.* Rods of Kraton 101 were extruded using an Instron Capillary Rheometer with a capillary $2''$ long and $0.125''$ in diameter. In order to produce a smooth, oriented rod, it was necessary to keep the capillary at a temperature below the T_g of polystyrene. Extrusion was stopped; the material was allowed to anneal in the capillary for 1/2 hour and was then forced out. The rod produced was then used for measurement of mechanical properties.

An ultrathin section of the rod was examined in the electron microscope. As can be seen in the micrograph (Figure 7), the polystyrene domains were in the form of parallel rods. The structure, however, was not nearly as well ordered as that reported by Keller and his co-workers.

In a second set of experiments, Kraton 101 was milled at 100°C and allowed to band on the mill. The milled sheet was highly birefringent with the orientation in the milling direction. This sheet was placed in a compression mold at 150°C under conditions that allowed almost no flow in the mold. The specimen was cooled in the mold before removal. The resultant sheet had highly direc-

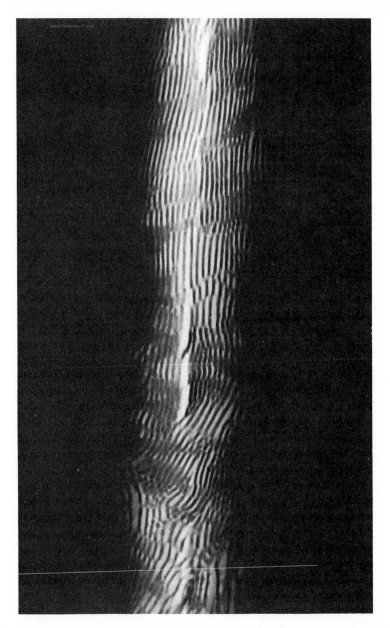

Figure 7. Electron micrograph of an ultrathin section of Kraton 101 extruded and annealed in an Instron Rheometer.

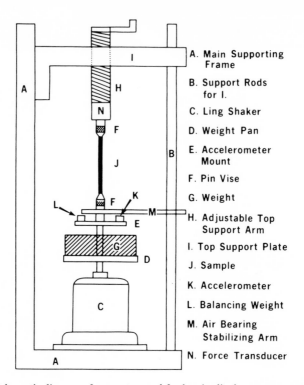

A. Main Supporting
   Frame

B. Support Rods
   for I.

C. Ling Shaker

D. Weight Pan

E. Accelerometer
   Mount

F. Pin Vise

G. Weight

H. Adjustable Top
   Support Arm

I. Top Support Plate

J. Sample

K. Accelerometer

L. Balancing Weight

M. Air Bearing
   Stabilizing Arm

N. Force Transducer

Figure 8. Schematic diagram of apparatus used for longitudinal resonance measurements.

tional properties: in one direction, the sample could be bent like a sheet of rubber, but in the transverse direction it bent more like a sheet of rigid plastic. Strips of this material were cut in both of these directions, and measurements were made in tension, torsion, and bending.

2. *Mechanical properties measurements.* The tensile modulus E_{11} of the rod-shaped specimen was measured from the longitudinal resonance frequency by using a small shaker to excite the sample. A schematic drawing of the apparatus used for these measurements is given in Figure 8.

The torsional modulus was measured using a compound torsion pendulum developed in our laboratory. This test makes use of the electronics of the B&K complex modulus apparatus [15]. A schematic drawing of the torsion pendulum is shown in Figure 9.

In both tests, moduli were measured as a function of preload, and damping was measured both from half-band width and from the logarithm of the decrement in free decay. Results of these measurements are given in Table I.

These same measurements were made on the sheet specimens in both longitudinal (orientation) and transverse directions. In addition, flexural measurements

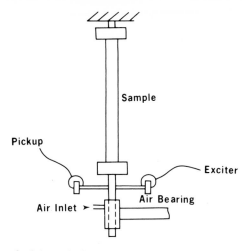

Figure 9. Schematic diagram of compound torsion pendulum.

were also made using the B&K complex modulus apparatus. A schematic drawing of this equipment is given in Figure 10. The flexural modulus was determined from the resonance frequency of a sample clamped at one end. The free end was driven by a magnetic transducer that attracted a small piece of magnetic foil attached to the free end. The displacement amplitude was determined by

TABLE I

Moduli of Extruded Kraton 101 Rod

Description of Test	G' (dynes/cm^2)	G'' (dynes/cm^2)
Sample #1		
Torsion modulus	5.4×10^7	3.9×10^6
After loading	6.5×10^7	4.2×10^6
	E' (dynes/cm^2)	E'' (dynes/cm^2)
Tensile modulus, initial	2.62×10^9	1.02×10^8
After loading to 150 psi	8.66×10^8	1.03×10^8
After unloading	8.30×10^8	$.99 \times 10^8$
After annealing	1.34×10^9	1.48×10^8
Sample #2		
Tensile modulus, initial	4.28×10^9	1.45×10^8
After loading to 150 psi	9.31×10^8	1.06×10^8
After unloading	8.07×10^8	0.97×10^8

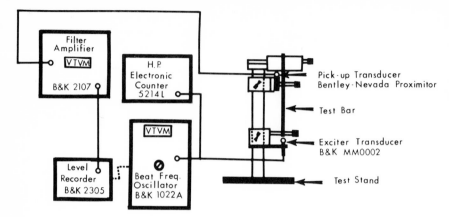

Figure 10. Schematic diagram of B&K complex modulus apparatus used for measuring flexural resonant modes.

means of a noncontacting proximity detector pick-up. Results of these experiments are given in Tables II and III.

3. *Discussion of results.* From Table I it can be seen that a Young's modulus in the extrusion direction (E_{11}) as high as 4.28×10^9 dynes/cm² can be obtained on the extruded and annealed rod samples. The ratio of the Young's modulus E_{11} to shear modulus G_{12} for Sample #1 is 49. A ratio greater than 80 would be expected for Sample #2. At loads of greater than 150 psi, there appears to be a yield process that results in a threefold decrease in tensile modulus. Annealing causes a partial recovery in that modulus. Torsion measurements after loading showed a small but significant increase in shear modulus.

As seen in Table II, the mill-oriented sheet had a ratio of 7:1 between Young's moduli in the milling and transverse directions, and a 20:1 ratio between the higher Young's modulus $(E_{11}) = 3.6 \times 10^9$ dynes/cm² and shear modulus $(G_{12}) = 1.72 \times 10^8$ dynes/cm². The sheet obviously was much less oriented than the rods, yet the Young's modulus in the milling direction was not much lower than that of the rod. Both the flexural and tensile resonance tests on strip speci-

TABLE II

Moduli of Mill-Oriented Kraton 101 Sheet

Test Method	Milling Direction		Transverse	
	E' or G' (dynes/cm²)	E'' or G'' (dynes/cm²)	E' or G' (dynes/cm²)	E'' or G'' (dynes/cm²)
Beam flexure	3.6×10^9	2.3×10^8	0.56×10^9	4.5×10^7
Tensile resonance	2.07×10^9	3.7×10^8	0.5×10^9	4.0×10^7
Torsion (shear)	1.72×10^8	8.7×10^6	1.8×10^8	8.9×10^6

TABLE III

Effect of Sample Length and Mode Number on Apparent Flexural Modulus
of Mill-Oriented Kraton Sheet

Mode No.	Length/Thickness	E' (dynes/cm^2)	E'' (dynes/cm^2)
1	86	3.5×10^9	2.0×10^8
3	86	3.6×10^9	1.7×10^8
5	86	3.4×10^9	1.8×10^8
1	55	3.7×10^7	
2	55	3.3×10^9	1.9×10^8
3	55	2.81×10^9	
1	28	2.83×10^9	1.6×10^8
2	28	2.42×10^9	1.40×10^8
3	28	2.20×10^9	

mens should measure Young's modulus. In the flexural test, the maximum strain level was less than 10^{-6}, and the layers most distant from the neutral plane make a disproportionately large contribution to the flexural stiffness. A discrepancy between the two values may be due to a low-strain yielding in the tensile experiment and/or a larger degree of orientation in the surface layers. Since the strain levels in torsion and flexural tests are comparable, these two values have been compared in the discussion below.

Table III shows another interesting observation made on these materials. Using the B&K complex modulus apparatus, measurements were made both as a function of sample length and as a function of frequency. Up to five resonance modes could be excited. Huang [16] has considered the problem of vibration of orthotropic beams. If the ratio $E_{11} : G_{12}$ is large, then there is considerable shear deformation in the beam. The effect will manifest itself as a decrease in the apparent shear modulus. The effect becomes more pronounced as the beam becomes shorter or the mode number higher. Effects of this type have been recently demonstrated for boron–epoxy composites [17]. As shown in Table III, the effect is quite pronounced for the sample oriented in the milling direction. The sample tested in the transverse direction did not display a significant shear effect.

The data presented above can be interpreted in terms of the Halpin–Tsai equations. The moduli were calculated for a system of polystyrene fibers uniaxially arranged in a polybutadiene matrix. The assumed moduli of polystyrene and polybutadiene and the calculated values for the composite are given in Table IV.

It can be seen that the Young's moduli of the rodlike samples before yielding is only half that of the calculated values. If we were to use Equation (4) to calculate moduli for a discontinuous fiber composite, the axial ratio would have to be 750 to achieve this modulus. The shear modulus is considerably higher than

TABLE IV

Calculated Moduli of Polystyrene–Polybutadiene Fiber Composites

A. Assumed Material Properties
 G (rubber) = 1 × 10^7 dynes/cm²
 E (rubber) = 3 × 10^7 dynes/cm²
 G (polystyrene) = 1.13 × 10^{10} dynes/cm²
 E (polystyrene) = 3.0 × 10^{10} dynes/cm²
 v (polystyrene) = 0.33
B. Calculated Properties of Orthotropic Composites Containing Continuous
 Polystyrene Rods
 G_{12} = 1.66 × 10^7 dynes/cm²
 E_{11} = 7.5 × 10^9 dynes/cm²
 E_{22} = 6.0 × 10^7 dynes/cm²
 v_{12} = 0.46
 E_{11}/E_{12} = 125
 E_{11}/G_{12} = 450
C. Calculated Moduli for Quasi-Isotropic System
 1. For Continuous Rods
  E(quasi-isotropic) = 2.87 × 10^9 dynes/cm²
  G(quasi-isotropic) = 0.95 × 10^9 dynes/cm²
 2. For Rods (L/D = 750)
  E(quasi-isotropic) = 1.50 × 10^9 dynes/cm²
  G(quasi-isotropic) = 0.50 × 10^9 dynes/cm²

the theoretical value: 5.4 × 10^7 vs. 1.66 × 10^7 dynes/cm². This discrepancy can be explained by a small amount of disordered structure. Using Halpin–Tsai equations, we find that 5 percent disorder would account for this discrepancy and yet would have little effect on the Young's modulus.

Both the micrographs and the mechanical property data indicate a considerable amount of structural imperfection in the system. At fairly low tensile stresses (e.g., 250 psi), large local stress concentrations could lead to rupture or yielding in the polystyrene domains. We believe that this is the process that accounts for the yielding observed in these experiments. The relative constancy of the loss moduli suggests that the dissipative processes originate mainly in the rubber phase and are not greatly influenced by the yielding process. Uniaxial orientation is not as closely approximated in the mill-oriented sheets as in the extruded rods.

The large decrease apparent in flexural modulus with vibrational mode or decreasing sample length suggests that the shear modulus G_{12} is small and that fibers are preferentially oriented parallel to the plane of the sheet. The data suggest a considerable amount of disorder in the system. We assume the system can be modeled as a laminate of orthotropic and quasi-isotropic plies. Again, to explain the Young's modulus in the principal direction, we would have to assume an axial ratio for the fibers of at least 750. If we assume 15 percent quasi-isotropic material, that is 15 percent disordered structure, we would closely approximate the observed moduli.

It can thus be seen that if we consider phase-separated block polymers to be microcomposite materials, then the well-established concepts and equations for predicting properties of composite materials can be a guide to understanding and interpretation of the mechanical behavior of block polymers. The model considered showed that even in the absence of structural regularity, this approach can be useful.

From composite theory, we see that by manipulating size, morphology, and properties of the phases, we can obtain a wide gradation of composite properties using simple starting materials. These properties are dependent upon the chemical structure of block polymers, which can be systematically varied by the technique of anionic polymerization.

Anionic Copolymerization in the Preparation of Block Polymers

Anionic polymerization offers one of the most powerful methods for the preparation of the block polymers of controlled structure. As is well known, the reactive species from such initiators as butyllithium, with certain monomers, are "long-lived" carbanions. Thus, one may polymerize a monomer A to completion, followed in sequence by polymerization of another monomer B to obtain a diblock polymer. The triblock A–B–A block polymers like Kraton can be prepared by sequential monomer addition either from a mono- or di-functional initiator or by the technique of chemically coupling two A–B molecules each containing a single active center. A number of kinetic and mechanistic details, however, must be considered in the preparation of block polymers of precisely defined structure from styrene with butadiene or isoprene by anionic polymerization. These have recently been discussed by several authors [18,19].

An aspect of anionic polymerization that makes it a very useful method for obtaining copolymers of controlled sequence length is the broad range of reactivity ratios that can be obtained in copolymerization of such monomers as styrene with the dienes by changing the nature of the solvent. This is described below in some detail for the anionic copolymerization of styrene and isoprene in different solvents, particularly some of our early studies [24], which have not been reported in detail elsewhere.

Anionic Copolymerization of Styrene and Isoprene

The results of Kelley and Tobolsky [20] given in Table V show the effect of solvents on the n-butyllithium-initiated copolymerization of isoprene and sty-

TABLE V

Percentage Styrene in Styrene–Isoprene Copolymers at Low Conversion

Solvent	Wt-% Styrene in Copolymers
Benzene	18 ± 1
Undiluted monomers	17 ± 1
Triethylamine	60 ± 3
Diethyl ether	68 ± 3
Tetrahydrofuran	80 ± 6

Note: Conditions: styrene–isoprene = 60/40 by weight (equimolar).

rene. As the Lewis base strength of the solvent toward organolithium increases, a greater amount of styrene is incorporated in the initially formed copolymer. This can be the result of an increased ionic character of the organolithium bond or a decrease in the association of the polymer$^-$ Li$^+$ species [21].

The polydiene microstructure is also altered by the solvent used in anionic polymerizations. Isoprene polymerized in more polar solvents contains larger amounts of the 3,4 and *trans*-1,4 structure. In hydrocarbon solvents, the polymer consists almost entirely of the *cis*-1,4 structure [22].

Tobolsky and Rogers [23] carried out an extensive study of the effect of different solvents on the structure of polyisoprenes prepared with *n*-butyllithium. Some of their results are shown in Figure 11. Of particular interest to us was the effect of diphenyl ether. Even in pure diphenyl ether, a relatively large amount of the 1,4 structure (i.e., *cis*- and *trans*-) is still obtained. From these data, it can be concluded that diphenyl ether acts as a rather weak Lewis base, probably as a result both of its steric hindrance in coordinating with the lithium cation and of the delocalization of the oxygen's electrons into the phenyl groups.

In the *n*-butyllithium-initiated copolymerization of isoprene and styrene in hydrocarbon solvents, isoprene is generally polymerized almost exclusively in the first part of the reaction, followed by polymerization of styrene to give a terminal block of polystyrene. In the first part of the reaction the rate approximates that of the homopolymerization of isoprene, whereas in the second part it is close to that of styrene. The first inflection point on the percent conversion-time curve corresponds to that point in the reaction where there is a rapid increase in the concentration of the poly(styryl lithium). At this point the reaction medium changes from being almost colorless to an orange-yellow color characteristic of poly(styryl lithium).

A conversion–time curve that we obtained for the polymerization in benzene of an isoprene–styrene mixture at an initial mole ratio of 0.774 is shown in Figure 12, where an inflection point occurs at 59 percent conversion. This and all subsequent polymerizations discussed were carried out at 30.0 ± 0.01°C. Samples were obtained throughout the course of the reaction, and the copolymer

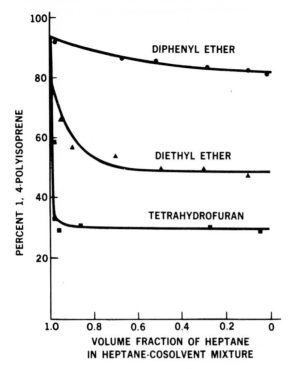

Figure 11. Effect of different solvents on polyisoprene microstructure.

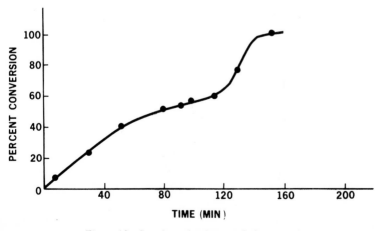

Figure 12. Copolymerization rate in benzene.

composition was determined from the infrared absorption corresponding to the C–H out-of-plane bending mode for styrene at 695 cm^{-1}. The precision of determining styrene content in this way was found to be ±2.5 percent. When it was assumed that $r_1 r_2 = 1$ for this copolymerization, a good linear plot was found when the data were treated according to the integrated form of Wall's equation [27]: $dM_1/dM_2 = r_1(M_1/M_2)$. Figure 13 gives a plot of log M_1/M_1^o versus log M_2/M_2^o, where M_1 refers to isoprene and M_2 to styrene. From these data, $r_1 = 7.7$ was obtained, which is in good agreement with values reported by others. For example, Korotkov and Rakova [25] reported values of r(isoprene) = 7.0 ± 0.6 and r(styrene) = 0.14 ± 0.02, while Spirin and co-workers found [26] for this reaction, in toluene, that r(isoprene) = 9.5 and r(styrene) = 0.25.

It is also possible to estimate, from the initial comonomer concentrations, the composition of the copolymer present at the first inflection point. This involves assuming that isoprene is completely consumed at the inflection point and using the measured extent of reaction at that point. For the experiment under consideration, we calculated that the copolymer present at the inflection point contained 43.1 percent styrene by weight. Analysis of the copolymer formed at

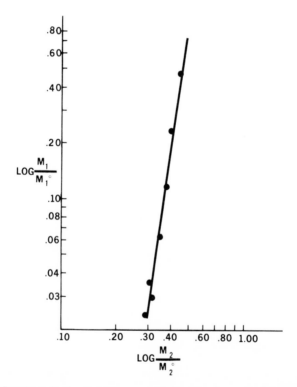

Figure 13. Copolymerization in benzene: reactivity ratios from copolymer composition.

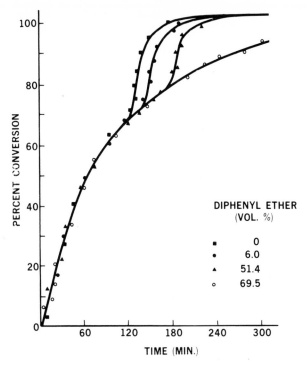

Figure 14. Effect of diphenyl ether on copolymerization rate.

this conversion showed that it contained 44.5 percent styrene by weight, which agrees within experimental error.

Copolymerization conversion curves for isoprene–styrene at 1 M total monomer concentration, 0.006 M in *n*-butyllithium initiator, and at different diphenyl ether concentrations are given in Figure 14. Rates were measured dilatometrically, and reactions were carried out in an all-glass high-vacuum (10^{-5}-10^{-6} Torr) system.

As the diphenyl ether concentration is increased, more styrene is polymerized in the initial part of the reaction. The inflection point is displaced to higher extents of reaction, and the length of the terminal polystyrene block decreases. Since all of the copolymerizations were at the same initial concentrations, these results indicate a change in the copolymerization reactivity ratios. Confirmation of this conclusion was sought by an attempt to determine the reactivity ratios for the copolymerization carried out in 51–54 volume-percent diphenyl ether. When it was assumed that $r_1 r_2 = 1$, an average value of r_1 (isoprene) = 3.4 was obtained. However, this value may be subject to a significant error due to the low molecular weight of the copolymer and to the difficulty of vacuum-drying the copolymer from diphenyl ether, which has a low vapor pressure. For these reasons, an alternative approach to determining reactivity ratios from anionic copolymerization rates alone was attempted. An example of this approach is outlined below for the copolymerization carried out at 51–54 volume-percent diphenyl ether.

Copolymerization rates were measured at different initial isoprene–styrene ratios at a 0.006 M n-butyllithium concentration. These results are given in Figure 15, where it is shown that the first inflection point is displaced to higher extents of reaction as the isoprene–styrene ratio is increased. It was established that the location of the inflection point is independent of the initiator concentration in the range 0.00095 M to 0.00835 M. Therefore, for a particular solvent system, the inflection point, dependent upon both the copolymerization reactivity ratios and the homopolymerization rates, is only dependent upon the initial comonomer ratio. The copolymer composition for each comonomer ratio was calculated from the extent of reaction at the inflection point, assuming that the isoprene concentration approached zero at this point. In order to calculate the reactivity ratio from these data, it was assumed that $r_1 r_2 = 1$ and that at the inflection point the M_1/M_2 ratio is a constant. The physical significance of this

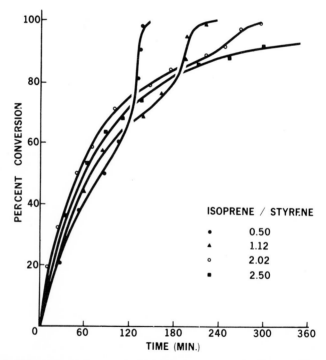

Figure 15. Effect of initial comonomer ratio on copolymerization rate (51–54 vol-% diphenyl ether).

last assumption is that the kinetics of the reaction are such that at this point in the reaction there is a tendency for the rate of reaction of the active centers with monomer to rapidly favor the formation of the poly(styryl lithium). Using Wall's integrated equation for copolymerization in the form $\log M_1^\circ/M_2^\circ = \log M_1/M_2 - (r_1 - 1) \log M_2/M_2^\circ$, $\log M_1^\circ/M_2^\circ$ versus $\log M_2/M_2^\circ$ was plotted to obtain the value of $(r_1 - 1)$. This approach was used since the value for M_1 at the inflection point, although small, is subject to a large error because of the manner in which it was calculated. The plot of the data for the reaction carried out in 51–54 volume-percent diphenyl ether is given in Figure 16. A value of r_1(isoprene) = 2.7 was obtained—not much different from that obtained by direct copolymer composition analysis.

As mentioned earlier, there is a change in the polyisoprene microstructure as the reaction is carried out in the presence of increased amounts of diphenyl ether. In Figure 17, the relationship between the 1,4 content of the polyisoprene and the amount of styrene incorporated prior to the inflection point is given for copolymerizations carried out at different diphenyl ether concentrations. It appears that the same factor leading to an increased 3,4 content of the polyisoprene facilitates a greater incorporation of styrene in the copolymer.

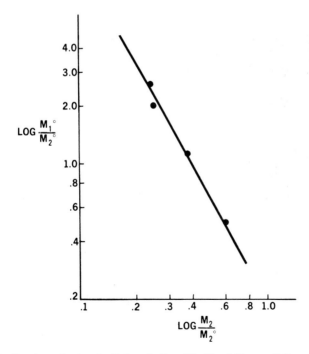

Figure 16. Copolymerization in diphenyl ether (51–54 vol-%): reactivity ratios from copolymerization rates.

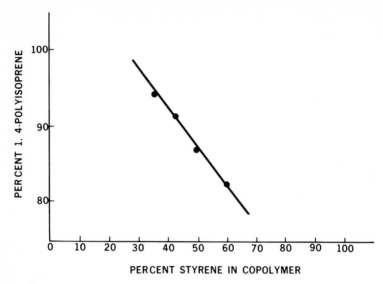

Figure 17. Relationship between microstructure of polyisoprene and copolymer composition.

Description of Structure of Block Copolymers of Styrene–Isoprene Prepared in Different Solvents

From the copolymerization studies discussed above, it is possible to provide a rather detailed picture of the structure of isoprene–styrene copolymers prepared using n-butyllithium in benzene and at various concentrations of diphenyl ether. In order to do this, the number average sequence length $\bar{n}$, for each of the co-monomers was calculated [28] at different degrees of conversion for r_1(isoprene) = 7.7, the value obtained in benzene, and r_1(isoprene) = 3, an approximate value for the reaction carried out in 51–54 volume-percent diphenyl ether. From the copolymerization rate data (see Figure 14), it is expected that the r_1 value for the reaction carried out in 6 volume-percent diphenyl ether would be only slightly less than that found in pure benzene, whereas the r_1 value for the reaction carried out in pure diphenyl ether, i.e., 69.5 volume-percent, would be less than 3 and would approach the value at which an inflection point is not readily detectable.

In the copolymer prepared in benzene (see Figure 18), there are three distinct structural regions. Initially, during the polymerization of the isoprene, the styrene run lengths interrupting the isoprene sequences are on the average of one monomer unit. Near the inflection point, the concentration of the comonomers is such that a close-to-random placement of the two monomers is obtained. After the inflection point, a terminal block is formed consisting almost entirely of

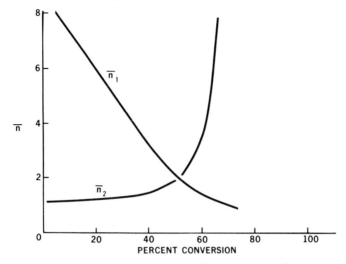

Figure 18. Average sequence length of isoprene $(\bar{n}_1)$ and styrene $(\bar{n}_2)$ with percent conversion for copolymerization in benzene, $r_1 = 7.7$

styrene. With increasing amounts of diphenyl ether present in the copolymerization (see Figure 19), the initial isoprene runs become smaller and the terminal styrene block becomes shorter. As a consequence of these changes, the middle copolymer portion is extended over a greater portion of the polymer chain.

Transition Behavior of Styrene–Isoprene Block Copolymers Prepared in Different Solvents

From the study described above, a series of styrene–isoprene copolymers of varying sequence lengths, with little change in the diene microstructure, were made available. These polymers were used in an attempt to determine the physical property changes resulting from a systematic variation in domain structure, with little change in chemical composition.

In order to study the mechanical behavior of the small amounts of sample prepared by the high-vacuum techniques, use was made of a coated-vibrating-reed procedure for measuring mechanical damping. With this method, the polymer is coated from a solvent onto a spring steel reed [30]. The coated reed is electromagnetically driven, and measurements are made of resonant frequency and damping as a function of temperature.

Three different solvent systems were used in this study. Benzene-cast films were used to differentiate the behavior of copolymers of differing structure, and solvent mixtures of benzene-heptane or tetrahydrofuran (THF)–methyl ethyl

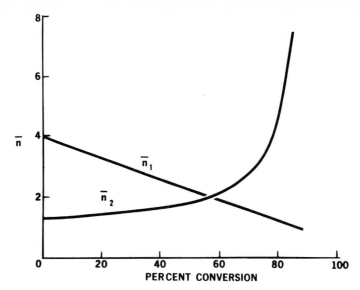

Figure 19. Average sequence length of isoprene ($\bar{n}_1$) and styrene ($\bar{n}_2$) with percent conversion for copolymerization in 51–54 vol-% diphenyl ether, $r_1 = 3$.

ketone (MEK) were investigated to enhance the separation of the styrene-rich or isoprene-rich components.

A 90/10 mixture of THF–MEK was chosen because both polystyrene and polyisoprene are soluble in THF, while only the styrene-rich portion should be soluble in MEK. Since tetrahydrofuran has the higher vapor pressure, it was expected to evaporate from solution first. Actual analysis of the solvent mixture during casting of the reeds substantiated this hypothesis. As the polymer solution becomes more concentrated, it contains a greater fraction of methyl ethyl ketone, and the isoprene-rich portion begins to separate from solution. Finally, as the methyl ethyl ketone is removed, the styrene-rich portion of the polymer begins to precipitate. When both solvents have been completely removed, the reed is coated with a film, which should consist of a polystyrene continuous phase with a polyisoprene-rich copolymer dispersed in it.

A 90/10 mixture of benzene–heptane was also chosen because heptane is a good solvent for the isoprene-rich portion of the block copolymer but not for the styrene-rich portion, while benzene, which is a good solvent for both parts of the copolymer, has a higher vapor pressure than heptane and would therefore be expected to evaporate from the reed first.

Comparison of Figures 20 and 21 demonstrates the effect of the chemical structure of the copolymers on their transition behavior. The copolymer prepared in diphenyl ether (see Figure 21), gives only a single, very broad transition over the temperature range from $-35°C$ to $+70°C$. We have demonstrated ki-

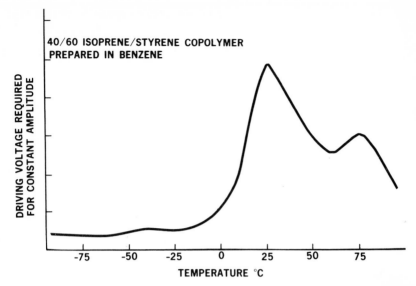

Figure 20. Damping vs. temperature for 40/60 isoprene–styrene copolymer prepared in benzene.

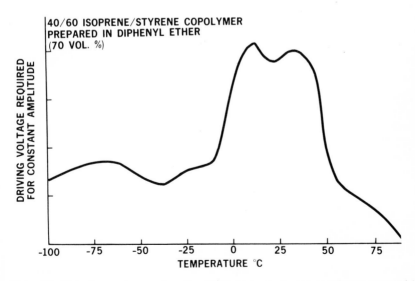

Figure 21. Damping vs. temperature for 40/60 isoprene–styrene copolymer prepared in diphenyl ether (70 vol-%).

netically that this polymer has a structure that is intermediate between that of a block polymer and that of a random copolymer.   However, the microblocks in this copolymer are sufficiently long to be of borderline miscibility.

For the copolymer prepared in benzene (see Figure 20), multiple transitions characteristic of a block polymer are observed. The transitions displayed are not completely assignable from the detailed structure of the polymer. Rather, they suggest that varying degrees of phase mixing also occur. The extent of phase mixing should depend on the solvent system used to cast the polymer films.

This concept was tested using a block polymer prepared in benzene containing 6 percent diphenyl ether. When cast from benzene, its transition behavior is that shown in Figure 22, which is very similar to the transition behavior of the polymer prepared in pure benzene. Use of the mixed solvent 90/10 THF–MEK enhances the phase separation of this polymer, as shown in Figure 23. On the other hand, when this polymer is cast from the 90/10 benzene–heptane mixture, phase separation, as shown in Figure 24, is much less complete than that cast from pure benzene. Thus, it is concluded that for a fixed copolymer composition the degree of phase separation of the components of a block polymer depends on which portion of the copolymer is made to phase-separate first.

The nature of the solvent used for coating the films, the casting temperature, and the rate of removal of solvents all may play a significant role in arriving at the final structure of the block polymer microcomposite. The reason for this is

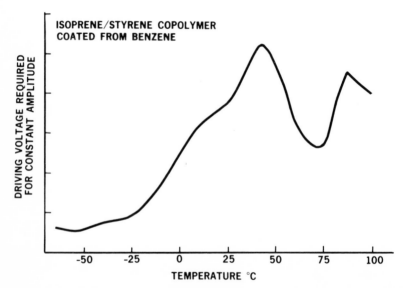

Figure 22. Damping vs. temperature for isoprene–styrene copolymer prepared in 6 vol-% diphenyl ether and coated from benzene.

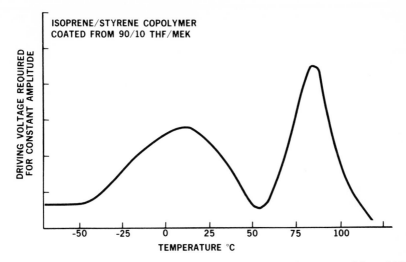

Figure 23. Damping vs. temperature for isoprene–styrene copolymer coated from 90/10 THF–MEK.

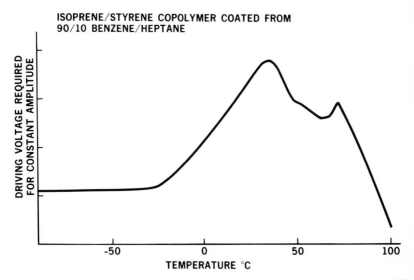

Figure 24. Damping vs. temperature for isoprene–styrene copolymer coated from 90/10 benzene–heptane.

that the final morphology of the phase-separated polymer is not necessarily an equilibrium state. This feature, rather than being a limitation of these materials, is their main attraction, since a particular polymer structure can be manipulated to form many different morphologies exhibiting different physical properties.

Phase Separation in Pure Block Polymers

Transition Behavior and Stress–Strain Properties

As discussed above, phase separation in microblock polymers (block polymers with short sequence length of the monomers in one of the blocks) is quite sensitive to the mode of the solvent treatment. Marked solvent effects can also be found in polymers in which the blocks are well defined. This may be seen, for instance, from some work on Kraton 101. Three solvent systems were used in this study: (1) a mixture of THF and MEK containing 90 percent THF by volume; (2) a mixture of benzene and heptane containing 90 percent benzene by volume; and (3) carbon tetrachloride. The basis for the choice of systems (1) and (2) has been discussed above. Carbon tetrachloride was chosen as a solvent because it has a solubility parameter of 8.63 $(cal/cc)^{1/2}$, which is between that of polystyrene (9.13) and that of polybutadiene (8.44).

A comparison of the transition properties of the three solution-cast specimens was made using the vibrating-reed apparatus. Figure 25 is a plot of damping vs. temperature for Kraton 101 deposited from the solvent systems employed in this study. All of these curves showed two distinct transitions: one in the vicinity of $-80°C$, close to the transition temperature of butadiene; the other in the vicinity of $100°C$. In each of these curves there is also a strong suggestion of a transition occurring at a temperature intermediate between those of the polystyrene and the polybutadiene transitions. Evidence for this intermediate transition is particularly clear in the case of the polymer cast from CCl_4. The way in which the modulus of each of these polymers changes with temperature is also of interest. The decrease in modulus on going through the glass transition was greatest for the sample cast from the benzene–heptane solution, intermediate for that cast from CCl_4, and least for that cast from the THF–MEK solution. These differences show up much more clearly in the room-temperature stress–strain curves (Figures 26 and 27). The most striking feature in the stress–strain curve of Kraton 101 is its "plasticlike" behavior, which is seen most dramatically in the case of specimens prepared by solution-casting from the THF–MEK solvent system. A yield point at about 3 percent elongation is clearly seen in the first extension cycle shown in Figure 26. During this drawing process, the force remains essentially constant. In the second elongation cycle, the deformation is homogeneous and the stress–strain curve is more like that of a typical vulcanized

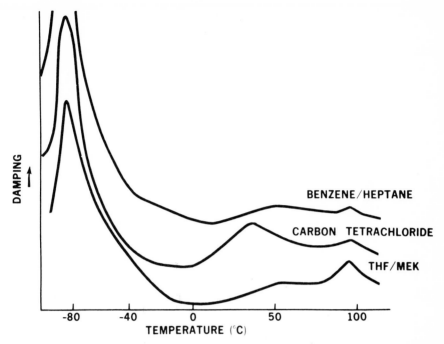

Figure 25. Damping vs. temperature for Kraton 101 on samples deposited from different solvent systems.

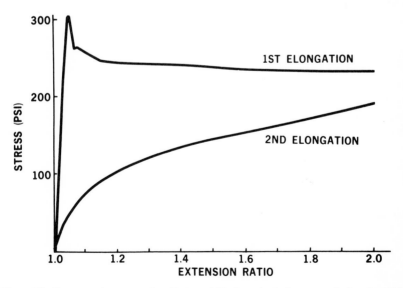

Figure 26. Stress–strain curve for Kraton 101 deposited from a solution in 90/10 THF–MEK.

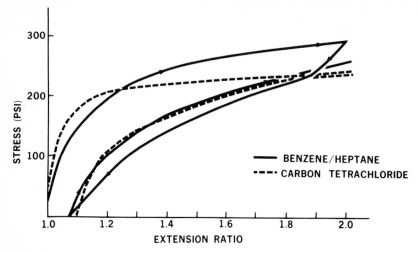

Figure 27. Stress–strain curves for Kraton 101 films prepared from solutions in 90/10 benzene–heptane and carbon tetrachloride.

elastomer. The existence of a yield point and the difference between the first and second elongation cycles suggest that some rigid structure has been broken during the drawing. The existence of such a rigid structure has been suggested by other workers [31,32].

Stress–strain curves for specimens cast from benzene–heptane solution are more "rubberlike" (Figure 27) than those cast from THF–MEK solutions. There is no yield point and, unlike the compression-molded and THF–MEK-cast specimens, no distinct drawing region. There is also considerably less difference between the mechanical behavior in the first extension cycle and that in the second extension cycle. The behavior of samples cast from CCl_4 is of particular interest. These samples appear to be more leathery than do samples cast from the other solvent systems. The stress-strain curves for both the first and the second extension cycles have higher initial slopes than do those for a sample cast from a benzene–heptane mixture. Although there is no additional set on subsequent cycling, recovery is slow, as is the stress relaxation on holding at constant length. This "leathery" response is consistent with the behavior of a polymeric material at a temperature near a damping maximum.

Some confirmation of this picture emerges from electron micrographs obtained on ultrathin films of cast polymer [2]. Although we now have evidence that micrographs of cast ultrathin films are not identical to those obtained from ultrathin sections cut from thicker films, these micrographs are still indicative of the types of differences that exist between these systems. We originally believed that the basic morphology of this system was a phase-separated sphere. Sufficient evidence now exists for rodlike structures, so that these micrographs should be reinterpreted. Suffice it to say that our micrographs suggested the existence of fairly well-separated particles in the system cast from the benzene-heptane system, and that there are frequent interconnections in the system cast from THF-MEK, with the inference of the existence of a continuous polystyrene phase. In the system cast from CCl_4, the phase boundaries are diffuse. In the system cast from a benzene–heptane solvent, the particles are much better separated and appear to be more like separated spheres.

From the above studies, we conclude that the morphology of cast block polymers is fixed in solution at some stage during the evaporation process. For this reason, the casting solvent is critical. It can change which phase is continuous or discontinuous, or it can lead to the existence of two continuous phases. There is also a good deal of evidence that phase separation is not necessarily complete and that this can lead to diffuse interphase boundaries and the existence of a third mixed phase.

Evidence for the intermediate transition has been found for a number of polymers and has shown up in both dynamic mechanical and thermal measurements of transition properties. Evidence for intermediate transitions and for even a great deal of change in the whole nature of the transition curve is, of course, most pronounced in the microblock polymers, where an almost continuous gradation of structural variations exists.

Thermodynamics of Phase Separation in Block Polymers

The transition behavior and mechanical properties of block polymers can be treated in terms of a simple thermodynamic model. High-molecular-weight polymers differing in chemical structure are normally incompatible, because mixing of polymers is usually endothermic, while the only driving force toward compatibility is the very small entropy of mixing. The important new consideration in block polymers is the restrictions that arise from the attachment of the blocks and from the small size of the domains. These restrictions cause a drastic reduction of the entropy of a polymer molecule in a phase-separated system as compared to that of a free chain.

Thermodynamic treatments of this phase-separation process have been given by a number of workers. Both Meier [34] and Kawai and his co-workers [33] have made predictions of the conditions for phase separation as well as the size and shape of the domain. We feel that Meier [34] has treated all the important features of the system. We have considered a rather simple model [35] based on similar considerations. We consider only the case in which the domains are lamellar, as shown schematically in Figure 28. Conformational entropy is calculated by considering the problem of random walk in a tetrahedral lattice. For this calculation, we assume (1) that there is complete phase separation between the two components and (2) that all junction points between the blocks occur at

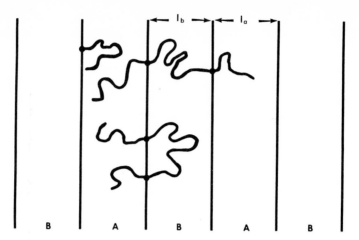

Figure 28. Model for phase-separated domain in triblock polymers.

the phase boundary. In an A–B–A triblock polymer, all B chains start at the phase boundary and terminate at either the same boundary or the opposite boundary of the B domain. All A chains start at the boundary and end in the domain interior. The length of the two domains are in the same ratio as the volume fraction of the block components. We further assume that the densities of these components are the same as those of the corresponding homopolymers.

Our object was to calculate the difference in free energy between the completely mixed system and that of the system separating into domains. The main part of the problem was to calculate the entropy terms that correspond to the constraints on chain configuration imposed by domain formation.

The details of the calculation are not important to this paper, but the following consequences of the entropy calculations should be pointed out.

1. Attachment of a chain to a wall restricts the number of configurations a chain can have. This restriction becomes less important as the chain becomes longer.

2. The size of the domain must be restricted since it cannot possibly exceed the extended chain length of the molecule for the diene domain and twice that length for the polystyrene domain. These highly extended domains would lead to a deficiency of material in the central part of the domain. This consideration has been treated by both Kawai and Meier. In our one-dimensional model, the domain never becomes sufficiently large for this to be important. The small size of the domain, however, is a serious restriction to the number of configurations a chain is allowed.

3. All chains must start at or near a domain boundary. This means that of all possible sites in the system, the chain junction (A) in Figure 28 can lie only on sites at or near the domain boundary giving rise to an entropy term that depends on the surface-to-volume ratio.

4. The second end of the diene chain must lie at or near one of two domain boundaries and must not pass through any boundary. This restriction is quite severe. This term also depends on surface-to-volume ratio.

Consequences 3 and 4 tend to limit the size of the domains. The balance between these and consequences 1 and 2 controls the equilibrium domain size.

On simple assumptions, the above simple model predicts that a completely mixed system can be more stable than the completely separated system. If, however, we relax the condition of complete separation at the interface, we get an increase in "interfacial energy," or heat of mixing, but receive in return a large bonus in terms of reduction of the placement entropy at the domain boundaries. Partial mixing in the vicinity of the phase boundary will always be favored over complete separation. We then see a region in which not only the chain junction is allowed to occur throughout a region, but the chain segments of both species can freely occur.

There is by now sufficient evidence for incomplete phase separation in many block polymers that this picture seems quite reasonable. The morphology of a solvent-cast polymer is not an equilibrium morphology but is established at some time during the evaporation process and represents some solution morphology. In solution, the above picture is somewhat modified. Solvent will partition between the two phases. The interfacial boundary would be expected to be more diffuse since the presence of a solvent decreases the number of polymer–polymer contacts and, hence, the enthalpy of mixing. Swelling of this domain, on the other hand, increases the number of conformations available to a chain.

The presence of the interfacial region should have another effect. If the interfacial region is below its glass temperature, segments of rubbery chains passing through the interfacial regions will act as cross-links and cause the system to have a much higher modulus than would otherwise be expected.

In some cases, e.g., some of the microblock polymers or Kraton 101 cast from CCl_4, we would expect that this would lead to leathery behavior. We would also expect that the temperature dependence would be greater than would be expected, e.g., from the WLF equation, and that there would be a fairly large change in modulus at intermediate temperatures. Such behavior is observed for many of the elastomer block polymers. In addition, even in cases in which there is no yield point, the stress–strain behavior deviates considerably from the predicted neo-Hookean or Mooney–Rivlin behavior. The initial modulus is very high and decreases rapidly with extension.

Summary and Conclusions

1. Block polymers have a unique capability for producing microcomposites of controlled morphology. Variations in both morphology and the nature of the phases can be effected by changes in the polymer structure and in the manner of preparing experimental samples.

2. In some cases, highly ordered structures can be produced. Composite theory can be helpful in explaining the mechanical properties of these polymers.

3. Microblock polymers produced by anionic copolymerization have a gradation of block structures. Phase separation in these systems can be to multiphase rather than two-phase structures in which there is poor definition of the domain boundary and considerable mixing of components. The properties of these polymers can be understood only by first understanding the details of the morphological arrangements. In its usual form, composite theory assumes sharp phase boundaries. For these more complicated systems, the theory can still give guidance to the expected behavior.

4. Transition peaks corresponding to those of the component blocks indicate phase separation. The nature of the solvent determines the extent of phase separation, which influences the breadth of the transition peaks. In the case of solvents that preferentially swell the dispersed phase, the domains are interconnected into rodlike ladder structures. This results in a yield point and the marked dependence of the stress–strain curve on the previous strain history.

5. Phase mixing at the interfacial boundary relaxes the entropy restrictions and makes a phase-separated structure in block polymers thermodynamically stable compared to either a completely mixed or a completely separated system.

Even though some of the materials produced so far have only limited applications, it can be projected with certainty that the concepts based on domain structure of polymers will play an ever-increasing role in the science of strong polymer materials.

References

1. Alfrey, T., Bohrer, J.J., and Mark, H., *Copolymerization,* New York: Interscience (1952).
2. Beecher, J.F., Marker, L., Bradford, R.D., and Aggarwal, S.L., "Morphology and Mechanical Behavior of Block Polymers," *J. Polym. Sci., Part C,* No. 26 (1969), 117.
3. McIntyre, D. and Campos-Lopez, E., "Small Angle X-ray Scattering Studies of Tri-Block Polymers," in *Block Polymers,* S.L. Aggarwal, ed., New York: Plenum Press (1970), 19.
4. Folkes, M.J. and Keller, A., "The Birefringence and Mechanical Properties of a 'Single Crystal' from a Three Block Copolymer," *Polymer,* 12 (1971), 222.
   Dlugosz, A., Keller, A., and Pedemonte, E., "Electron Microscope Evidence of a Macroscopic 'Single Crystal' from a Three Block Copolymer," *Kolloid Z.-Z. Polym.,* 242 (1970), 1125.
5. LaFlair, R.T., "Structure Morphology and Properties of Block Copolymers," *XXIIIrd International Congress of Pure and Applied Chemistry,* Vol. 8, London: Butterworths (1971), 195.
6. Hashin, Z. and Rosen, B.W., "The Elastic Moduli of Fiber-Reinforced Materials," *J. Appl. Mech.,* 31 (1964), 233; "Theory of Composite Materials," *Proceedings of the Fifth Symposium on Naval Structural Mechanics,* F.W. Wendt, H. Leibowitz, and N. Perrone, eds., New York: Pergamon Press (1970), 201; and "Viscoelastic Behavior of Heterogeneous Media," *J. Appl. Mech.,* 32 (1965), 630.

7. Hermans, J.J., "Elastic Properties of Fiber Reinforced Materials When Fibers are Aligned," *Proc. Roy. Acad., Amsterdam,* B70 (1967), 1.

Chow, T.S. and Hermans, J.J., "The Elastic Constants of Fiber Reinforced Materials," *J. Compos. Mater.,* 3 (1969), 382.

8. Hill, R., "Theory of Mechanical Properties of Fibre-Strengthened Materials: I Elastic Behavior," *J. Mech. Phys. Solids,* 12 (1964), 199.

9. Kerner, E.H., "The Electrical Conductivity of Composite Media," *Proc. Phys. Soc., London, Sect. B,* 69 (1956), 802.

10. Keskkula, H., "Styrene Polymers (Plastics)," *Encyclopedia of Polymer Science and Technology,* Vol. 13, 396.

11. Roff, W.J. and Scott, V.R., *Fibers, Films, Plastics, and Rubbers,* London: Butterworths (1971), 371.

12. Ashton, J.E., Halpin, J.C., and Petit, H.P., *Primer on Composite Materials Analysis,* Stamford, Conn.: Technomic (1969), 77.

13. Lekhnitskii, S.G., *Anisotropic Plates,* S.W. Tsai and T. Cheron, trans., New York: Gordon and Breach (1968), 295.

14. Pagano, N.J. and Tsai, S.W., "Invariant Properties of Composite Materials," *Progress in Materials Science Series,* Vol. 1, Stamford, Conn.: Technomic (1968), 233.

15. Oberst, H., "Schwingungsdämpfende Kunstoffe aus optimal eingestellten Polymeren," *Kolloid-Z.,* 216–17 (1967), 64.

16. Huang, T.C., "The Effect of Rotary Inertia and of Shear Deformation on the Frequency and Normal Mode Equations of Uniform Beams in the Simple End Conditions," *Trans. ASME,* E, 28 (1961), 579.

17. Dudek, T.J., "Young's and Shear Moduli of Unidirectional Composites by a Resonant Beam Method," *J. Compos. Mater.,* 4 (1970), 232.

18. Morton, M., "Problems in the Synthesis of Block Polymers by Anionic Mechanism," in *Block Polymers,* S.L. Aggarwal, ed., New York, Plenum Press (1970), 1.

19. Prud-homme, J. and Bywater, S., "Block Copolymers of Styrene and Isoprene: Experimental Design and Product Analysis," in *Block Polymers,* S.L. Aggarwal, ed., New York: Plenum Press (1970), 11.

20. Kelley, D.J. and Tobolsky, A.V., "Anionic Copolymerization of Isoprene and Styrene. I," *J. Amer. Chem. Soc.,* 81 (1959), 1597.

21. Morton, M. and Ells, F.R., "Absolute Rates in Anionic Copolymerization," *J. Polym. Sci.,* 61 (1962), 25.

Johnson, A.F. and Worsfold, D.J., "Anionic Copolymerization of Styrene and Butadiene," *Makromol. Chem.,* 85 (1965), 273.

22. Stavely, F.W. and co-workers, "Coral Rubber-A *cis*-1,4-Polyisoprene," *Ind. Eng. Chem.,* 48 (1956), 778.

23. Tobolsky, A.V. and Rogers, C.E., "Isoprene Polymerization by Organometallic Compounds. II," *J. Polym. Sci.,* 40 (1959), 73.

24. Livigni, R.A., Marker, L.F., Shkapenko, G., and Aggarwal, S.L., "Structure and Transition Behavior of Isoprene–Styrene Copolymers of Different Sequence Length," paper presented at American Chemical Society, Division of Rubber Chemistry, Symposium on Structure and Properties of Elastomers, Montreal, Canada, May 1967.

25. Korotkov, A.A. and Rakova, G.V., "The Copolymerization of Isoprene and Styrene with Butyllithium," *Polym. Sci. USSR,* 3 (1962), 990.

26. Spirin, Yu. L., Polyakov, D.K., Gantmakber, A.R., and Medvedev, S.S., "The Homopolymerization and Copolymerization of Isoprene Initiated by Ethyllithium," *Polym. Sci. USSR,* 3 (1962), 233.

27. Wall, F.T., "The Structure of Copolymers. II," *J. Amer. Chem. Soc.,* 66 (1944), 2050.

28. Alfrey, T., Bohrer, J.J., and Marks, H., *Copolymerization,* New York: Interscience (1952), 134.
29. Marker, L., Aggarwal, S.L., Holle, R.E., and Kollar, W.L., "Time–Temperature Dependence of Mechanical Properties of Polymers and Copolymers of Propylene Oxide," *Kaut. Gummi Kunst.,* 19 (1966), 33.
30. Atkinson, E.B. and Eagling, R.F., "Some Applications of Dynamic Elastic Measurements in Polymer Systems," in *The Physical Properties of Polymers,* S.C.I. monograph No. 5, New York: Macmillan (1959), 197.
31. Fisher, E. and Henderson, J.F., "Effect of Temperature on Styrene Butadiene Block Copolymers," paper presented at the American Chemical Society, Division of Rubber Chemistry, Symposium on Structure and Properties of Elastomers, Montreal, Canada, May 1967.
32. Childers, C.W. and Kraus, G., "Properties of Random and Block Copolymers of Butadiene and Styrene. III. Three Sequence Styrene–Butadiene–Styrene Block Polymers," *Rubber Chem. Tech.,* 40 (1967), 1183.
33. Inoue, T., Soen, T., Hashimoto, T., and Kawai, H., "Studies on Domain Formation of the A-B-Type Block Copolymer from Its Solutions. Ternary Polymer Blend of the Styrene–Isoprene Block Copolymer with Polystyrene and Polyisoprene," *Macromolecules,* 3 (1970), 87.
34. Meier, D.J., "Theory of Block Copolymers. I. Domain Formation in A–B Block Copolymers," *J. Polym. Sci., Part C,* No. 26 (1969), 81.
35. Marker, L., "Phase Equilibria and Transition Behavior of Block Polymers—A Simple Model," *Polym. Prepr., Amer. Chem. Soc., Div. Polym. Chem.,* 10, no. 2 (1969), 524.

10. Multilayer Thermoplastic Sheets and Films

TURNER ALFREY, JR.
The Dow Chemical Company
Midland, Michigan

ABSTRACT

Multilayer thermoplastic sheets and films can be produced by coextrusion of two or more molten polymers. Some multilayer systems exhibit interesting mechanical effects ("mutual interlayer reinforcement") or optical effects (strong reflection of certain wavelengths). One method of generating a large number of layers utilizes an annular die with rotating surfaces. The layer distributions in films produced by such a die agree well with predicted distributions calculated from flow analysis.

Melt coextrusion has become an important commercial fabrication process for manufacturing layered thermoplastic composites. This type of process makes it possible to combine the desirable properties of several polymers without the expense incurred when several separately prepared films are laminated together.

Some properties of multilayer composites are essentially "additive." For example, the gas permeability behavior of some multilayer films can be calculated from the properties of the individual plastic materials and the layer geometry by "adding resistances in series" [1]. In other cases, synergistic interactions between the layers are observed. In some composites made up of alternating brittle and ductile layers, crack propagation is blocked, forcing the normally brittle layers to undergo ductile deformation [1]. Finally, interesting optical effects are observed when the individual layers are about 1,000 Å thick. Reflectivity spectra are determined by layer geometry [2].

One method of producing such multilayer structures utilizes an annular flow channel between rotatable die boundaries. The individual thermoplastic polymers enter this die through radial feedports; the combination of axial flow down the channel and rotation of the die boundaries leads to the generation of a large number of layers. The measured layer distributions agree well with predicted distributions calculated from flow analysis [3].

Mutual Mechanical Reinforcement in Multilayer Films

It is well known that multicomponent laminates often exhibit mechanical properties superior to those of the individual materials that constitute the separate layers. Some sword makers of early times hammered down alternate layers of hard and soft steel and obtained blades that took a fine cutting edge and yet were strong and tough [4,5]. Today, multilayer plastic film laminates are manufactured in wide variety; in each case the composite, multilayer laminate exhibits some property or combination of properties that cannot be matched by any one of the constituent materials.

This section deals with a specific particular mechanical effect sometimes encountered in such laminates—but an effect that is not limited to any particular film compositions. Thin layers of a high-modulus, low-elongation material are alternately sandwiched between thin (adhering) layers of high-elongation material. The film is tested to failure in tension, and it is observed that the high-modulus material undergoes a large ductile deformation before failing, which is in striking contrast to the free-film tensile behavior of this same material.

The high-elongation layers operate to prevent the propagation of transverse cracks across the brittle layers. With crack propagation so blocked, the stress in the brittle layers can reach the ductile yield point, and all layers can stretch out, together, to large elongations. Such a composite film has both a high modulus and a high elongation. It has a much higher work-to-break than would be exhibited by any of the individual layers.

Mutual interlayer reinforcement effects have been described in detail for mylar–aluminum–mylar composites, polyethylene–polystyrene–polyethylene composites, and 125-layer polystyrene–polypropylene films [1].

Optics of Iridescent Multilayered Films

Multilayer films containing hundreds of layers can exhibit vivid optical effects if the component polymers have different refractive indexes and if the layer distributions are properly adjusted [2]. The intensity of such reflections depends on the mismatch in refractive index.

Consider, for example, a 201-layer film containing 101 odd layers and 100 even layers. Let n_{odd} and d_{odd} represent the refractive index and layer thickness of the odd layers, and n_{even} and d_{even} those of the even layers. The wavelength of the strong first-order reflection (at normal incidence) is given by

$$\lambda_I = 2(n_{odd}d_{odd} + n_{even}d_{even}). \tag{1}$$

For example, if the thicknesses of the odd and even layers are set at 700 Å and 746.5 Å, and the refractive indexes are 1.6 and 1.5, respectively, then

$$n_{odd}d_{odd} = 1.6 \times 700 = 1,120,$$

$$n_{even}d_{even} = 1.5 \times 746.5 = 1,120,$$

$$\lambda_I = 2(1,120 + 1,120) = 4,480 \text{ Å}.$$

Strong reflection will occur in the wavelength region $\lambda = 4,480 \pm 100$ Å.

Such films, with *uniform* optical thickness, can exhibit strong reflectance, but only in a narrow wavelength band. To develop vivid reflection colors, it is necessary to have high reflectance over a broader wavelength range. The development of broad reflection bands requires the use of *varying* layer thickness through the multilayer film. Consider, for example, a 500-layer film, where $n_{odd} = 1.6$ and $n_{even} = 1.5$, and the layer thicknesses vary as follows with the number of a given layer (m):

$$d_{odd} = 700 + 0.5 \, m, \tag{2}$$

$$d_{even} = 746.5 + 0.533 \, m. \tag{3}$$

Such a film would reflect strongly over the wavelength range from 4,500 Å to 6,000 Å.

In addition to the first-order reflection, λ_I, higher-order reflections ($\lambda_{II}, \lambda_{III}, \lambda_{IV}, \ldots \lambda_M \ldots$) are observed at shorter wavelengths:

$$\lambda_I = 2 \, (n_{odd}d_{odd} + n_{even}d_{even}) \tag{4a}$$

$$\lambda_{II} = \frac{2}{2}(n_{odd}d_{odd} + n_{even}d_{even}) \tag{4b}$$

$$\lambda_{III} = \frac{2}{3}(n_{odd}d_{odd} + n_{even}d_{even}) \tag{4c}$$

$$\lambda_{IV} = \frac{2}{4}(n_{odd}d_{odd} + n_{even}d_{even}) \tag{4d}$$

$$\lambda_M = \frac{2}{M}(n_{odd}d_{odd} + n_{even}d_{even}), \tag{4e}$$

where λ_M is the Mth-order reflection.

Although the spectral *locations* of reflections of various orders are determined entirely by the *sum* of optical thicknesses of the two adjacent layers, the relative intensities of the various orders are strongly dependent upon the *ratio* between the two optical thicknesses. Let

$$f = \left(\frac{n_{odd}d_{odd}}{n_{odd}d_{odd} + n_{even}d_{even}}\right) \tag{5}$$

Equal optical thickness ($f = 0.5$) results in complete vanishing of all even orders and very strong reflections at all odd orders. If the odd layer comprises 1/4 of the total optical thickness (or 3/4), the second-order reflection is very strong, but the fourth-order vanishes. Higher-order reflection behaviors corresponding to several other values of f are listed below:

> $f = 0.15$ to 0.2; first four orders present; III strong
> $f = 0.33$; III vanishes
> $f = 0.4$; I and IV strong.

To summarize, thin multilayer plastic films can exhibit iridescence resulting from constructive reinforcement of radiation reflected from the multilayer interfaces. The character of the reflection spectrum depends on the refractive index mismatch of the two phases, the number of layers, and the thicknesses of the various layers. Narrow reflection bands can be developed in films where all odd layers are identical and all even layers are identical. The peak reflectance (first-order) occurs at the wavelength λ_I:

$$\lambda_I = 2(n_{odd}d_{odd} + n_{even}d_{even}).$$

Broad reflectance spectra require some systematic *variation* of layer thicknesses through the film. Relative intensities of the various higher-order reflections depend on the ratios of optical thickness of the adjacent layers.

Coextrusion of Multilayered Laminates

One method of producing large numbers of layers in a controllable and predictable manner is by means of an annular die channel with rotatable surfaces [3].

The individual components of the composite are pumped through a feedport system and into the die in alternating layers that extend radially across the annular gap. Simultaneous rotation of the die members, i.e., the inner mandrel and outer ring, deforms the layers into long thin spirals around the annulus. In effect, the number of layers is multiplied because of the increased interfacial surface area. The number of layers and their thicknesses are determined by the dimensions of the annulus, the number of feedports for each phase, the extrusion rate, and the rotational speed of the die mandrel and ring relative to the feedports. There are four basic layer patterns that may be generated by four modes of die rotation. These patterns are designated as Cases I, II, III, and IV.

Case I: inner die mandrel rotating; outer ring stationary. Layers are comparatively thick near the stationary outer ring and are progressively thinner and more numerous approaching the rotating mandrel. At the surface of the mandrel a large number of thin layers is generated.

Case II: inner die mandrel stationary; outer ring rotating. Layer pattern is essentially the inverse of Case I; i.e., the layers are thinnest near the outer die ring.

Case III: both inner and outer die members rotating at the same speeds and in the same directions. The layer pattern consists of curved open-end loops with very thick layers in the center of the annulus that become thinner near the boundaries.

Case IV: inner and outer die members counter-rotating at equal speeds but in opposite directions. The maximum number of layers is generated by this mode of operation. The layer distribution is very nearly symmetrical about the center of the annulus. The layers are thickest at the center of the annulus and become progressively thinner near each boundary.

The distribution of layer thicknesses in the extruded tube can be calculated by using a simplified Newtonian fluid model. If both polymers were Newtonian fluids of equal viscosity, the rotational velocity would vary with radial position as below:

$$v_\theta(r) = \frac{T}{4\pi\eta L}\left(\frac{r}{r_o^2} - \frac{1}{r}\right),$$ (6)

and the axial velocity would be given by

$$v_z(r) = \frac{\Delta P}{4\eta L}\left[\frac{(R_1^2 - R_2^2)}{\ln\frac{R_2}{R_1}}\ln r + r^2 + \frac{(R_2^2 \ln R_1 - R_1^2 \ln R_2)}{\ln\frac{R_2}{R_1}}\right]$$ (7)

The helix angle ϕ for the helical path followed by a moving particle would be given by

$$\phi(r) = \cot^{-1}\frac{v_\theta(r)}{v_z(r)},$$ (8)

and the total angle θ developed at a downstream position L would be

$$\theta = \frac{v_\theta(r)}{v_z(r)} \cdot \frac{L}{r}.$$ (9)

This permits the calculation of the layer pattern at the die exit. As the tubular extrudate leaves the die, the axial velocity distribution changes from roughly parabolic to uniform, and the individual layer thicknesses are thereby altered. This die-exit adjustment can be calculated from material balance considerations. Layer distributions calculated from the above simplified Newtonian model agree well with experimentally determined distributions.

Since the two polymers are actually highly non-Newtonian melts with different flow curves, some comment is in order regarding the above experimental agreement with predictions derived for Newtonian fluids of equal viscosity. First, consider the case where the two polymers are non-Newtonian but identical. The velocity field involves only two components, which are functions of r only: $v_z(r)$ and $v_\theta(r)$. Consequently, the rate-of-strain tensor has only two nonvanishing components: $\dot{\epsilon}_{rz}$ and $\dot{\epsilon}_{r\theta}$. At every point a local coordinate frame exists in which the rate-of-strain tensor has only *one* nonvanishing component. In this coordinate frame, the deviatoric stress tensor has only one component. The rate of strain can be calculated from this stress component and the steady-state fluid-flow law; transformation back to the (r, θ, z) coordinate frame yields $\dot{\epsilon}_{r\theta}$ and $\dot{\epsilon}_{rz}$, and then $v_z(r)$ and $v_\theta(r)$ can be obtained by integration. The Newtonian model yields a good value for the *ratio* $(\dot{\epsilon}_{rz}/\dot{\epsilon}_{r\theta})$ even when the fluid is actually non-Newtonian. If, for example, the polymer is actually a power-law fluid, the Newtonian approximation is grossly in error with regard to $v_z(r)$ and $v_\theta(r)$, but is reasonably accurate with regard to the *ratio* (v_z/v_θ). Since it is this ratio that determines the helix angle of flow and, hence, the downstream striation pattern, the simple Newtonian model provides a useful estimate of the quantities of interest here.

Next consider the case where the two polymers are *different*. The rates of shear, $\dot{\epsilon}_{r\theta}$ and $\dot{\epsilon}_{rz}$, will alternate between high and low values from one layer to another. The equal viscosity model ignores this completely. However, it is the ratio between the two *velocity* components that determines the helix angle and the striation pattern. In a system with *many layers,* the velocity components $v_z(r)$ and $v_\theta(r)$ approach smooth curves even when the rates of shear are strongly discontinuous. For the quantities of interest here, the simplified analysis again yields useful predictions.

Summary

Multilayer thermoplastic laminates containing up to hundreds of layers can be generated in a controllable and predictable manner by coextrusion through a rotating annular die channel. Multilayer composites can exhibit interesting mechanical and optical effects.

References

1. Schrenk, W.J. and Alfrey, T., "Some Physical Properties of Multilayered Films," *Polym. Eng. Sci.,* 9, no. 6 (1969), 393.
2. Alfrey, T., Gurnee, E.F., and Schrenk, W.J., "Physical Optics of Iridescent Multilayered Plastic Films," *Polym. Eng. Sci.,* 9, no. 6 (1969), 400.
3. Schrenk, W.J. and Alfrey, T., "Coextrusion of Blown Multilayer Plastic Films," *Amer. Chem. Soc., Div. Org. Coatings Plast. Chem. Prepr.,* 32, no. 1 (1972), 205.

4. Slayter, G., "Composites," *Sci. Amer.,* 206, no. 1 (1962), 124.
5. Smith, C.S., "Decorative Etching and the Science of Metals," *Endeavour,* 16, no. 64 (1957), 199.
6. Schrenk, W.J., Cleereman, K.J., and Alfrey, T., "Continuous Mixing of Very Viscous Fluids in an Annular Channel," *SPE (Soc. Plast. Eng.) Trans.,* 3, no. 3 (1963), 192.

MECHANICAL AND PHYSICAL PROPERTIES

MODERATOR: ARTHUR V. TOBOLSKY
Frick Chemical Laboratory
Princeton University
Princeton, New Jersey

11. Infrared Studies of Polyurethane Block Polymers

STUART L. COOPER AND ROBERT W. SEYMOUR

University of Wisconsin
Madison, Wisconsin

ABSTRACT

Extensive mechanical and thermal testing has led researchers to postulate that the properties of block polymers may be ascribed to the existence of a domain morphology. Such microstructure has been confirmed for hydrocarbon block polymers by electron microscopy. Transmission electron micrographs of segmented polyether- and polyester-urethane elastomers have been obtained and show a similar domain structure in these materials. Infrared techniques have allowed studies of the extent and nature of the hydrogen-bonding possibilities in polyurethane block polymers. Exploration of various deformation and temperature histories shows that hydrogen bonding is extensive and not easily disrupted. A measure of the orientation of polymer chains in the different domains has been obtained using infrared dichroism. The soft segments are readily oriented by an applied stress but return to the unoriented state when the stress is removed. The hard domains, which are composed of aromatic urethane linkages, orient to much the same extent as the soft domains, but have a tendency to remain in the oriented conformation after removal of the applied stress. The set of the aromatic domains provides a molecular interpretation for the phenomenon of stress hysteresis in these systems. Further correlations between the results of the infrared experiments, measurements of the dynamic mechanical spectra, and differential scanning calorimetry of segmented polyurethanes are presented.

The unusual mechanical properties and morphology of block polymers have been extensively studied in recent years and have led to the application of some of these materials as thermoplastic elastomers. Thermoplastic elastomers are composed of incompatible segments having glass-transition temperatures above and below the use temperature. The segments above their glass temperature are designated the soft segments; those below T_g, the hard segments. In polyurethane block polymers, the soft segments are generally polyethers or polyesters, one to five thousand in molecular weight, and the hard segments are formed from the extension of a diisocyanate (often aromatic) with a low-molecular-weight diol. The polar urethane segments provide the physical cross-linking responsible for the block polymer's cohesion and elastic properties. Since no covalent cross-links are present, the material may be processed as a thermoplastic at temperatures above the softening transition temperature of the hard segment.

The existence of microphase separation caused by clustering of at least some of the hard and soft segments into separate domains has been well established [1–3]. The mechanical properties of these materials reflect their heterophase nature and are characterized by an enhanced rubbery modulus and extensibilities of several hundred percent [4, 5]. Block polymer systems exhibit the major glass or melting transitions of the two components as well as various secondary loss mechanisms [6].

The typical polyurethane is extensively hydrogen-bonded [7], the donor being the NH group of the urethane linkage. The hydrogen-bond acceptor may be in either the hard urethane segment (the carbonyl of the urethane group) or the soft segment (an ester carbonyl or ether oxygen). The relative amounts of the two types of hydrogen bonds are determined by the degree of microphase separation, with increased phase separation favoring interurethane hydrogen bonds.

Hydrogen-bonding considerations have been shown to influence morphological features such as chain ordering in partially crystalline polyurethanes [8, 9]. The importance of hydrogen bonds in determining mechanical properties is less clear, however. The hetero-bonded solid-state concept [10] predicts that such secondary bonding will play a major role in dynamic mechanical transition behavior. Studies of hydrogen-bonded polymers have generally concluded, though, that the presence of such bonds does not necessarily enhance mechanical properties [11–13]. Other authors are of the opposite opinion, however [14].

Much attention has been directed toward an elucidation of the domain morphology in polyurethane block polymers, as their physical properties may be directly ascribed to the presence of this two-phase microstructure. The behavior of the elastomer under deformation is a function of both the domain size and the degree of order within the domains [8, 9, 15]. The original morphology may be altered through stretching to high elongations [8, 9, 15], annealing [16], and annealing under strain (heat setting) [8, 17]. The importance of such studies is apparent, because an understanding of how the morphological features may be altered is fundamental to the development of structure–property relationships.

Experimental Approaches

A wide variety of experimental techniques have been used in the study of polyurethanes, several of which are briefly reviewed. Only those methods relating to solid-state structure–property relations are considered.

Dynamic Mechanical Properties

Dynamic mechanical properties of polyurethane elastomers have been studied by Kimura et al. [18] and Huh and Cooper [6]. Figure 1 summarizes the results of Huh and Cooper for a polyester urethane as a function of segment length. The low-temperature γ peak (about $-100°C$) is assigned to a localized motion in the methylene sequences. The α peaks are associated with soft-segment micro-Brownian motion (α_a) and crystallite melting (α_c). The δ peak is attributed primarily to hard-segment micro-Brownian motion, and δ' to dis-ordering of the hard domains. A most interesting result of this work is the apparent effect of domain size and segment length on the position and intensity of the α_a relaxation. As expected, the magnitude of this dispersion decreases as the soft segment becomes more crystalline. Surprisingly, however, the temperature of the peak also decreases with increasing segment length. Since the longer segments are expected to produce better-ordered and larger domains, some

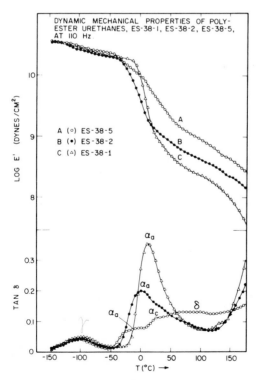

Figure 1. Dynamic mechanical properties of polyester urethanes as a function of segment length (segment length increases ES-38-1 to ES-38-5).

soft segments can exist in regions well removed from hard-domain interactions so that their motion may be relatively unrestricted by the hard domains.

Thermal Analysis

Various thermoanalytical techniques such as DTA, DSC, thermomechanical analysis, and thermal expansion measurements have been employed in the study of morphology and intermolecular bonding in polyurethane block polymers. The well-known thermal lability of the hydrogen bond has led to the interpretation of thermoanalytical data primarily in terms of hydrogen-bond disruption. Specifically, a DTA or DSC endotherm in the region of 80°C has been ascribed to the dissociation of the urethane–soft-segment hydrogen bonds, while an endotherm around 150-170°C is related to the breakup of interurethane hydrogen bonds [19-21]. In addition, a higher-temperature melting endotherm from microcrystalline hard segments is observed in materials having longer urethane segments. DSC curves for a polyether urethane are shown in Figure 2 [16]. It is seen that the behavior is strongly dependent upon thermal history in a fashion not entirely consistent with previous interpretations. The endotherm

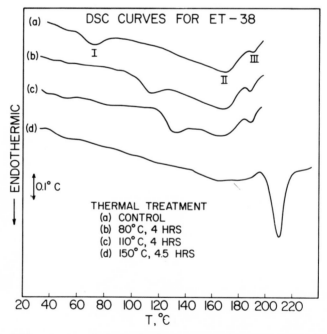

Figure 2. DSC curves for a hydrogen-bonded polyether urethane as a function of thermal history.

ascribed to hard–soft segment interaction (I) may be moved continuously up the scale in temperature by annealing until it merges with the presumed interurethane hydrogen bond dissociation endotherm (II). Severe annealing can lead to a material showing only a single micro crystalline peak (III) in the DSC. Materials of low diisocyanate content (and thus short hard segments) are incapable of crystallization and show only the merging of the I and II peaks under annealing.

The annealing studies cast doubt on the interpretation of the I and II peaks in terms of hydrogen-bond dissociation for several reasons. First, it seems unlikely that the strength of the hydrogen bonds themselves should be affected by annealing, though this could be implied by the data. Second, the I peak is continuous with the II peak in the sense that the two regions may be combined into a single endotherm by annealing. Interpretation in terms of hydrogen bonds would also require the conclusion that there be no bond dissociation below the III temperature in a well-annealed sample, as there are no lower temperature peaks. Infrared evidence that this is not true is discussed below.

Recent DSC studies of polyurethane block polymers having no capability for hydrogen bonding confirm that the DSC endotherms do not result from disruption of specific secondary-bond interactions (Figure 3). The curves shown are for a polyether urethane with a piperazine hard segment. The remarkable similarity to the curves of Figure 2 strongly suggests that relating the DSC endotherms to hydrogen-bond dissociation may be seriously questioned. It rather appears that all three endotherms are morphological in origin. The III

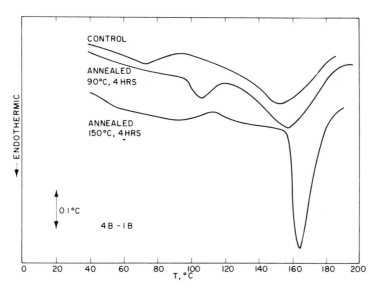

Figure 3. DSC curves for a non-hydrogen-bonded polyether urethane as a function of thermal history.

peak may be assigned to relatively well-ordered microcrystalline aromatic polyurethane segments, the numbers and perfection of which are determined by the segment length and thermal history. The I and II endotherms represent disordering of hard segments with relatively short-range order that may be improved in a continuous manner by annealing. The appearance of two peaks and the broad temperature region over which they occur are partially the result of the wide distribution of hard-segment lengths. Clusters of shorter hard segments, perhaps of only one diisocyanate unit, give rise to the lower-temperature endotherm, while the longer segments disorder at higher temperatures. This short-range order may be reorganized into microcrystalline regions if the hard segments are sufficiently long, though this transformation occurs in a discontinuous manner [16]. Improvement of short-range ordering is still possible in materials not capable of microcrystallization.

X-Ray Studies

Depending upon the segment length, polyurethane block polymers may exhibit crystallinity in both the macroglyclol and aromatic urethane segments. The d spacings observed in the copolymer can generally be related to spacings found in the constituent homopolymers. However, the primary use of X-ray techniques with polyurethanes has not been in crystal structure analysis but rather in orientation studies.

These investigations, on materials characterized as having paracrystalline [8, 9] or crystalline [15] hard segments, concluded that initially the hard domains orient so that the chain axis is preferentially transverse to the stretch direction. Further elongation causes the hard segments to slip past one another, breaking up the original domain structure, and becoming increasingly oriented into the direction of stretch.

Small-angle X-ray scattering has been used to detect the presence of microphase separation [3] and qualitative alterations of it [13]. This technique has not been used with polyurethanes to the extent that it has with other block polymer systems in which supradomain ordering has been studied [22, 25].

Light Scattering

The existence of supradomain ordering in some polyurethanes has been detected with low-angle light scattering by Wilkes [26]. A spherulitic texture was found, with spherulite size increasing with hard-segment length. Sepherulitic structures have also been reported by Kimura et al. [15]. Deformation studies showed the texture to be quite stable and apparently unaffected by the domain disruption, with increasing strain implied by chain-orientation studies.

Infrared Techniques

The use of infrared spectroscopy depends on the presence of spectral features that can be identified with particular aspects of the morphology, intermolecular associations, or molecular structure. While the infrared spectra of polyurethane block polymers are quite complex, they do satisfy these requirements. Typical spectra for polyurethanes based on an aromatic (diphenylmethane diisocyanate) hard segment are shown in Figure 4. Three regions are of interest: the NH stretch absorptions (3,500–3,200 cm^{-1}), the CH$_2$ stretch absorptions (3,000–2,700 cm^{-1}), and the carbonyl vibrations (1,750–1,650 cm^{-1}).

Participation in hydrogen bonding decreases the frequency of the NH vibration and increases its intensity, making this absorption very useful in the study of hydrogen-bond effects. The peak is located at about 3,320 cm^{-1} in the spectra of Figure 4, which is characteristic of hydrogen-bonded NH groups. Nonassociated NH absorbs around 3,450 cm^{-1}, with a significantly smaller absorption coefficient [27]. Changes in hydrogen bonding can thus be followed, in principle, by frequency or intensity measurements.

The carbonyl absorption is also of potential use in hydrogen-bond studies. In practice this is limited to polyether urethanes, because the presence of two types of carbonyl groups in polyester urethanes gives only a broad, unresolved peak (Figure 4). Frequency shifts on hydrogen bonding also occur for the carbonyl vibration, though the effect is much smaller for the acceptor than the donor (30 cm^{-1} vs. 125 cm^{-1}). The free and bonded peaks are resolved in the polyether urethane spectrum of Figure 4, and their relative absorbances can be

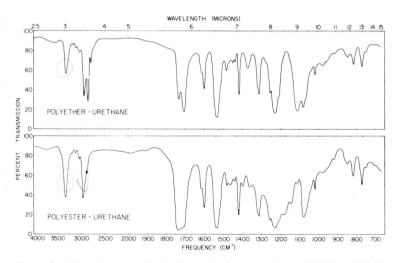

Figure 4. Infrared spectra of polyether and polyester urethanes ET-38 and ES-38.

used as a measure of the degree to which this group participates in hydrogen bonding.

Molecular orientation can also be studied spectroscopically, using the technique of infrared dichroism. It is of particular interest in block polymer systems when the dichroic behavior of bands characteristic of the individual blocks can be studied. This is possible if there are infrared-active groups peculiar to each segment that absorb in regions free from other bands. A quantitative description of segmental orientation also requires a reasonably accurate knowledge of the transition-moment directions for the vibrations of interest. In polyurethanes, the NH group is characteristic solely of the hard segment and may be conveniently used to study hard-block orientation. The CH_2 group can generally be used for soft-segment orientation, although methylene groups are also generally present at small concentrations in the hard segment. The transition-moment direction for both of these vibrations can be taken as $90°$ [28].

Results and Discussion

Scope of the Investigations

As has been discussed, application of infrared techniques to polyurethane block polymers allows direct study of the behavior of the hydrogen bonds and characterization of segmental orientation. The hydrogen-bonding studies have been conducted with a dual aim. The first is an understanding of the effects of chemical composition, deformation, and temperature on the hydrogen-bond network. In addition to this, it was anticipated that this knowledge would allow assessment of the role of hydrogen bonding in interpretation of other properties, such as dynamic mechanical spectra, calorimetry, and orientation behavior. The orientation behavior itself is studied because of its potential of providing a molecular interpretation relating morphology and extent of mechanical interaction between the phases to macroscopic mechanical properties. One example of this approach is the question of understanding the mechanism of stress hysteresis or Mullins effect observed in polyurethane block polymers [5].

The materials used in these studies were polyether (polytetramethylene oxide) and polyester (polybutylene adipate) urethanes having a hard segment formed from p, p'-diphenylmethane diisocyanate (MDI) chain extended with butanediol. The details of their chemical structure have been published previously [6,28]. These materials are referred to in this paper by the soft-segment type, ES (polyester) or ET (polyether), with weight-percent diisocyanate and soft-segment molecular weight in thousands by a code such as ES-38-1. If no soft-segment molecular-weight designation is made, it is 1,000. The DSC of a polymer containing a polyether soft segment and piperazine–butanediol hard segment has also been studied. It is from a series of polymers prepared by Dr. L. L. Harrell, Jr., which have verying segment lengths and segment polydispersity [29], and is designated 4B-1B. It contains 22 weight-percent piperazine, with broad size distributions in both segments.

Hydrogen-Bonding Studies

Examination of the NH region of polyether and polyester urethanes of a broad range of compositions (24 to 38 percent MDI) reveals that all are extensively hydrogen-bonded at room temperature. A free NH peak is never resolved, though in some cases a slight shoulder on the high-frequency side of the bonded NH peak may be observed. A variation in the type of hydrogen bond present is found with composition, however. As was previously pointed out, the hydrogen-bond acceptor may be in either the soft or the hard segment. It is reasonable to expect that relatively more bonds will be formed with soft-segment acceptors as the amount of hard segment is decreased. This distribution of bonds between the two acceptors may be studied in polyether urethanes by using the relative intensities of the free and bonded carbonyl peaks. A "hydrogen-bonding index," R, has been defined as the ratio of the absorbance of the

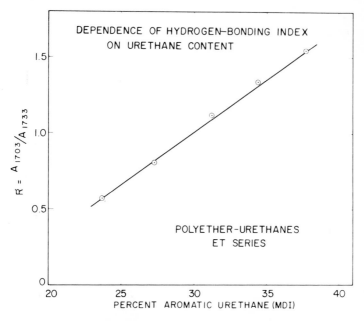

Figure 5. Hydrogen-bonding index as a function of hard-segment content for polyether urethanes.

bonded peak $(1,703 \text{ cm}^{-1})$ to the free $(1,733 \text{ cm}^{-1})$. This is plotted as a function of hard-segment content in Figure 5. The anticipated decrease in degree of interurethane hydrogen bonding (of which R is a measure) with decreasing hard-segment content is observed. The lessened interurethane bonding implies less complete phase separation at low urethane content.

The degree of phase separation can thus be increased by increasing the relative amount of hard segment. In addition, it may be increased at constant composition by increasing the length of the blocks. For example, doubling the segment length at the 38 percent MDI level (i.e., going from ET-38-1 to ET-38-2) increases R from 1.55 to 2.22.

Uniaxial stretching does not result in significant changes in the extent of hydrogen bonding, as is clear from Figure 6. The NH peak height (normalized by the asymmetric CH_2 stretch absorbance to account for thickness changes) remains nearly constant up to 300 percent strain. While it is apparent that any large-scale chain movement must be accompanied by disruption of hydrogen bonds, they are quickly reformed.

Up to 200 percent strain, there is also no change in the value of R (Figure 7). The slight decrease in R at higher strains, coupled with the constancy of the overall degree of hydrogen bonding, may indicate some preference for formation of hard–soft bonds as a result of large deformation. The effect is small, however, and it is reasonable to conclude that stretching causes no appreciable changes in the extent or type of hydrogen bonding.

The remaining question concerns the thermal lability of the hydrogen bonds. This has been investigated by studying the temperature dependence of infrared absorption for the NH vibration. The NH region as a function of temperature

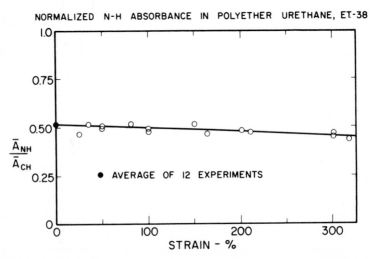

Figure 6. Normalized NH absorbance as a function of elongation for ET-38-1.

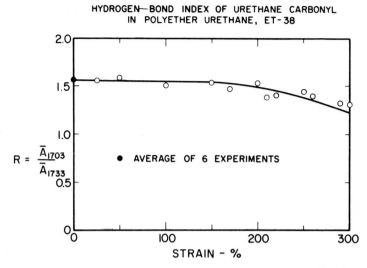

Figure 7. Hydrogen-bonding index as a function of elongation for ET-38-1.

for ET-38-1 is shown in Figure 8. At 25°C virtually all of the NH groups are hydrogen-bonded, giving the single peak at 3,320 cm^{-1}. As the temperature is raised, a high-frequency shoulder develops and the overall intensity diminishes. Nonbonded NH groups are known to absorb at a higher frequency and with much lower intensity, so that the latter effect can be used to monitor hydrogen-bond disruption. To characterize this dissociation quantitatively, the area of the total NH peak has been measured as a function of temperature. Because of the intensity decrease with bond disruption, the area should decrease as the number of free NH groups increases. This method was chosen because it was impossible to resolve adequately the free and bonded NH absorptions. Typical data analyzed in this way for ES-38-1 and ET-38-1 are shown in Figure 9.

We attribute the change in slope of the absorbance curve to the onset of hydrogen-bond dissociation, which occurs at the glass-transition temperature of the hard segments. The rigidity of the hard segments below their T_g would be expected to restrict hydrogen-bond disruption. As the temperature is raised and sufficient mobility is attained, secondary-bond dissociation can occur more readily, perhaps even according to the equilibrium law commonly applied to low-molecular-weight compounds and solutions. This concept is embodied in the theoretical treatment of Wolkenstein and Ptitsyn [30]. Slope discontinuities in absorbance–temperature curves have previously been used to detect transitions in several polymers, including those without specific secondary-bond interactions. A theoretical justification for this effect has been developed [31].

This analysis demonstrates that hydrogen-bond dissociation does occur, increasing steadily as the temperature is raised above the hard-segment T_g. It is

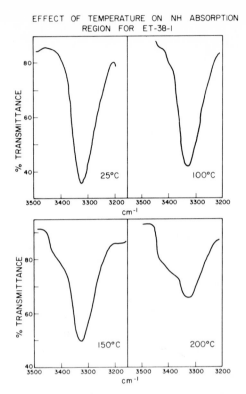

Figure 8. Effect of temperature on NH absorption region for ET-38-1.

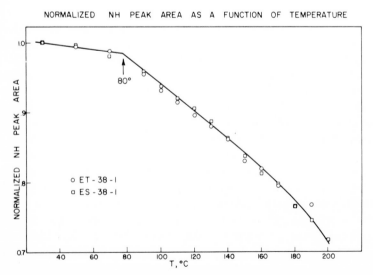

Figure 9. Integrated NH absorbance as a function of temperature for ES-38-1 and ET-38-1.

important to note, however, that there is still significant hydrogen bonding at 200°C (Figure 8) and that the thermal behavior of the hydrogen bonds is insensitive to the morphological details. No annealing effects are seen in the IR data, whereas they dominate the DSC (Figure 2).

Orientation Studies

Infrared dichroism is a powerful tool for the study of orientation in polyurethanes because of the presence of bands characteristic of the individual phases [28]. The data presented here are analyzed in terms of the simple axial orientation function f [32] for the hard segments (NH) and soft segments (CH).

Figure 10 displays the orientation functions for both the hard and soft segments in ES-38-1. The orientation of both segments is of similar magnitude, but the CH orientation at low-strain levels is greater than the NH orientation, with the two functions crossing at about 150 percent strain.

Residual orientation functions for the two segments are plotted in Figure 11. Residual orientation functions are obtained by stretching the sample to a given elongation, allowing it to relax for five minutes at zero load, then measuring the orientation. The abscissa in Figure 11 is the strain level to which the samples were extended prior to relaxation; hence, it is labeled prestrain. Although both soft and hard segments orient to nearly the same degree on initial straining, when relaxed the hard segments (NH orientation) exhibit substantially larger orienta-

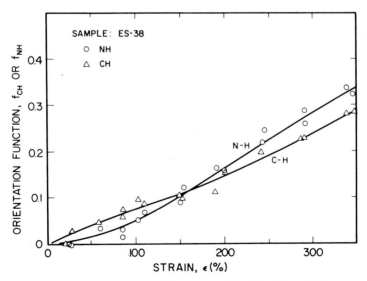

Figure 10. Orientation functions vs. strain for ES-38-1 at 28°C.

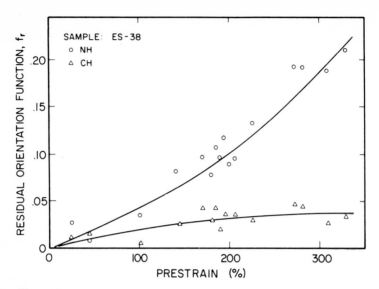

Figure 11. Residual orientation functions vs. prestrain for ES-38-1 at 28°C.

tion functions than the soft segments. Both ET-38-1 and ES-38-1 show this behavior. Since the elastomers creep considerably (65–70 percent following straining to 300 percent), it might be thought that the residual orientation is due to the residual strain. This could be true in part, but it cannot explain the entire residual hard-segment orientation, for at 70 percent initial strain the orientation function for NH alignment is only about 0.07. Moreover, if creep were entirely responsible, the residual orientation functions for the two phases should be approximately equal, but lower than those in Figure 10. It appears that the differences in the residual orientation functions are due to different retractive stresses acting on the separate domains.

To investigate further the connection between strain history and residual orientation, orientation functions were measured on films prestrained to 200 percent elongation. The results of these experiments are shown in Figure 12, along with comparable data for nonprestrained samples.

For ES-38-1 the orientation of the soft segments is relatively insensitive to prestrain, while the orientation of the hard segments is very dependent upon strain history. At strain levels lower than that of the prestrain (200 percent), the hard segments are oriented to a higher degree than is observed for nonprestrained samples. However, at extensions above the prestrain level, the NH orientation functions are essentially independent of prestrain. The high residual orientation of the hard domains is related to the stress-softening observed in the repeated stress–strain experiments of segmented polyurethanes that have been reported previously [5].

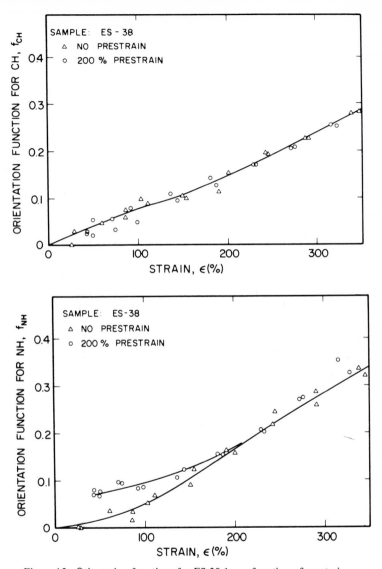

Figure 12. Orientation functions for ES-38-1 as a function of prestrain.

By monitoring the NH and CH infrared stretching absorptions, it has thus been shown that the orientation of different backbone segments depends strongly on strain history. On initial straining, both segments orient similarly; but after relaxation, chains containing the NH group remain oriented to a substantially higher degree than those containing the CH functional group. If the sample is again strained, at strain levels lower than those previously applied, the NH-containing chains are more highly oriented than during initial extension. The orientation of backbones containing CH groups is unaffected by the prestrain. The distinct difference between the orientation functions for the two types of chains provides further evidence that the different types of block polymer segments reside in different domains, which react somewhat independently to the application and removal of stress.

Orientation as a function of temperature has also been studied. Data at selected temperatures up to 120°C are shown in Figure 13. Both the NH and CH orientation functions at a given strain level go through a maximum with respect to temperature. The orientation functions increase gradually with increasing temperature up to about 90°C, above which they decrease rapidly with temperature. This behavior is demonstrated in Figure 14, which shows the temperature dependence of orientation functions measured at 200 percent extension. The maximum in orientability may be understood as the result of two thermally activated processes occurring simultaneously. The first is an ordering process due to softening of the domains, which allows greater orientation under the applied strain. This causes an increase in the orientation function.

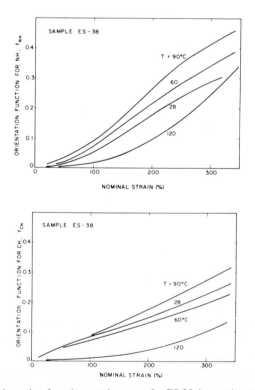

Figure 13. Orientation function–strain curves for ES-38-1 at various temperatures.

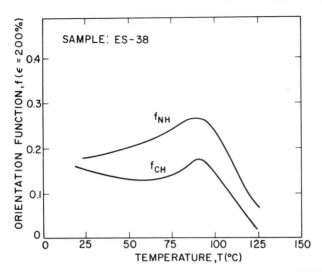

Figure 14. Orientation functions at 200% strain vs. temperature for ES-38-1.

The second process is disruptive, as the forces responsible for domain cohesion decrease with increasing temperature. This effect may be identified with the glass transition of the hard segments and the onset of hydrogen-bond break-up. It becomes dominant above 90°C, where the orientation function begins to drop with temperature.

Conclusions

The data discussed have clearly established the value of infrared techniques in the study of block polymers such as the polyurethanes. Under favorable conditions, knowledge of the behavior of secondary-bond interactions, transitions, and specific segment orientation may be obtained. Much of this information is available only by implication from other techniques. The relationship observed between mechanical hysteresis and residual segment orientation illustrates the potential of combining infrared techniques with macroscopic property measurements to provide a molecular understanding of bulk physical behavior.

Several important conclusions concerning the role of hydrogen bonding can also be made on the basis of the infrared evidence discussed. Initial interpretations of DSC and other thermal scanning experiments were in terms of hydrogen-bond dissociation [19, 20]. DSC annealing studies [16] combined with direct measurements of hydrogen-bond disruption by infrared methods strongly suggest that this interpretation is incorrect. It appears that the DSC endotherms arise from the loss of various degrees of long- and short-range order. This is interesting both from the viewpoint of the role of secondary bonds and for an understanding of ordering in amorphous polymers. Different degrees of short-range order, which may be continuously improved by annealing, can exist simultaneously. The thermal stability of the hydrogen bonds is insensitive to the degree of ordering present and is affected primarily by the glass-transition temperature of the hard segments. Above this temperature bond dissociation may be governed by an equilibrium law, as suggested in some theories [30], though we have no evidence for or against this idea.

The dynamic mechanical spectra of these polyurethanes show few, if any, characteristics that need be ascribed to hydrogen bonding [6]. It seems that chain mobility controls secondary-bond dissociation, rather than the opposite. This suggests that, at least in these systems, the hetero-bonded solid-state concept [10], which predicts that such secondary bonding would play a major role in transition behavior, does not apply.

Hydrogen bonds should not be regarded as "tie-down" points, particularly at temperatures above the hard-segment T_g; nor should they be expected necessarily to enhance mechanical properties. Such lack of enhancement has been noted by several workers [11-13], primarily in long-time relaxation studies where the hydrogen-bond interchange rate is appreciably greater than the experimental time scale. The unusually good mechanical properties of segmented polyurethanes must be ascribed to the incompatibility of the segments resulting in phase separation, rather than the presence of hydrogen bonds per se.

The presence of hydrogen bonds does serve to increase the overall cohesion of the materials, however, because these bonds are stronger and more directional than other intermolecular forces. This is reflected in the well-known fact that hydrogen-bonded compounds have higher melting points than non-hydrogen-bonded analogs [27]. In a similar manner, glass-transition temperatures in polymers may be elevated by the presence of hydrogen bonds [11, 14].

Hydrogen-bonding requirements may also be important in determining inter-chain alignments in highly ordered hard-segment regions. Such an approach has been successfully used by Bonart [8, 9] in the assignment of hard-segment X-ray reflections, and by MacKnight [33] in discussing the melting behavior of non-segmented polyurethanes.

Overall, the presence and behavior of the hydrogen bonds have much less effect on the mechanical properties of these polyurethanes than has been generally thought. Interpretation of structure–property relationships should be based more on morphological factors than on those dealing with specific intermolecular associations.

Summary

Examples of the variety of infrared measurements that may be useful in polymer property studies have been discussed. Infrared results may often be successfully correlated or contrasted with those from other techniques to provide insight into morphology, mechanical properties, and thermal behavior. Of particular utility in the study of polyurethanes have been studies of secondary bonding, transitions and molecular orientation as functions of temperature, deformation, thermal history, and composition.

Acknowledgment

The authors wish to thank Dr. Edward A. Collins of the B. F. Goodrich Chemical Company for supplying the MDI-based polyurethanes and Dr. L. L. Harrell, Jr., of E. I. du Pont de Nemours and Company for the piperazine-based elastomer. We are also grateful to the National Science Foundation for support of this research through Grants GK-4554 and GH-31747.

References

1. Estes, G.M., Cooper, S.L., and Tobolsky, A.V., "Block Polymers and Related Heterophase Elastomers," *J. Macromol. Sci., Rev. Macromol. Sci.,* C4, (1970), 313.
2. Koutsky, J.A., Hien, N.V., and Cooper, S.L., "Some Results on Electron Microscope Investigations of Polyether-Urethane and Polyester-Urethane Block Copolymers," *J. Polym. Sci., Part B,* 8 (1970), 353.
3. Clough, S.B., Schneider, N.S., and King, A.O., "Small-Angle X-Ray Scattering from Polyurethane Elastomers," *J. Macromol. Sci., Phys.,* B2 (1968), 641.
4. Cooper, S.L. and Tobolsky, A.V., "Viscoelastic Behavior of Segmented Elastomers," *Textile Res. J.,* 36 (1966), 800.
5. Cooper, S.L., Huh, D.S., and Morris, W.J., "Stress-Softening in Cross-linked Block Copolymer Elastomers," *Ind. Eng. Chem., Prod. Res. Develop.,* 7, no. 4 (1968), 248.
6. Huh, D.S. and Cooper, S.L., "Dynamic Mechanical Properties of Polyurethane Block Polymers," *Polym. Eng. Sci.,* 11 (1971), 369.
7. Seymour, R.W., Estes, G.M., and Cooper, S.L., "Infrared Studies of Segmented Polyurethane Elastomers. I. Hydrogen Bonding," *Macromolecules,* 3 (1970), 579.
8. Bonart, R., "X-Ray Investigations Concerning the Physical Structure of Cross-Linking in Segmented Urethane Elastomers," *J. Macromol. Sci., Phys.,* B2, no. 1 (1968), 115.
9. Bonart, R., Morbitzer, L., and Hentze, G., "X-Ray Investigations Concerning the Physical Structure of Cross-Linking in Urethane Elastomers. II. Butanediol as Chain Extender," *J. Macromol. Sci., Phys.,* B3, no. 2 (1969), 337.
10. Andrews, R.D. and Hammack, T.J., "The Theoretical Interpretation of Dynamic Mechanical Loss Spectra and Transition Temperatures," *J. Polym. Sci., Part B,* 3 (1965), 655.
11. Otocka, E.P. and Eirich, F.R., "Hydrogen Bonding in Butadiene Copolymers. I. Network Formation," *J. Polym. Sci., Part A-2,* 6 (1968), 895.
12. Fitzgerald, W.E. and Nielsen, L.E., "Viscoelastic Properties of the Salts of Some Polymeric Acids," *Proc. Roy. Soc., London, Ser. A,* A282 (1964), 137.
13. Tobolsky, A.V. and Shen, M.C., "The Effect of Hydrogen Bonds on the Viscoelastic Properties of Amorphous Polymer Networks," *J. Phys. Chem.,* 67 (1963), 1886.

14. Ogura, K. and Sobue, H., "Infrared Spectroscopic Approaches to Polymer Transitions. II. The Transition of Hydrogen Bonding in Styrene–Methacrylic Acid (St–MAA) Copolymers," *Polym. J.,* 3 (1972), 153.

15. Kimura, I., Ishihara, H., Ono, H., Yoshihara, N., and Kawai, H., "Morphology and Deformation Mechanism of Segmented Polyurethanes in Relation to Spherulitic Crystalline Texture," *Macromol. Prepr., XXIII Int. Union Pure Appl. Chem.,* 1 (1971), 525.

16. Seymour, R.W. and Cooper, S.L., "DSC Studies of Polyurethane Block Polymers," *J. Polym. Sci., Part B,* 9 (1971), 689.

17. Huh, D.S., "Morphology and Physical Properties of Polyurethane Block Polymers," unpublished Ph.D. dissertation, University of Wisconsin, 1971.

18. Matsumoto, K., Kimura, I., Saito, K., and Ono, H., "Property and Structure of Segmented Polyurethane. I. Linear Viscoelasticity," *Rep. Prog. Polym. Phys. Jap.,* 12 (1970), 279.

19. Clough, S.B. and Schneider, N.S., "Structural Studies on Urethane Elastomers," *J. Macromol. Sci., Phys.,* B2 (1968), 553.

20. Miller, G.W. and Saunders, J.H., "Thermal Analyses of Polymers. III. Influence of Isocyanate Structure on the Molecular Interactions in Segmented Polyurethanes," *J. Polym. Sci., Part A-1,* 8 (1970), 1923.

21. Vrouenraets, C.M.F., "DTA Studies of Polyurethane–UREA Block Copolymers," *Polym. Prepr., Amer. Chem. Soc., Div. Polym. Chem.,* 13, no. 1 (1971), 529.

22. McIntyre, D. and Campos-Lopez, E., "The Macrolattice of a Triblock Polymer," *Macromolecules,* 3 (1970), 322.

23. Brown, D.S., Fulcher, K.U., and Wetton, R.E., "Evidence of Simple Cubic Geometry from Small-Angle X-Ray Scattering in Three-Block Polymers," *J. Polym. Sci., Part B,* 8 (1970), 659.

24. Keller, A., Pedemonte, E., and Willmouth, F.M., "Macro Lattice from Segregated Amorphous Phases of a Three Block Copolymer," *Kolloid Z.-Z. Polym.,* 238 (1970), 385.

25. Kämpf, G., Krömer, H., and Hoffman, M., "Long-Range Order of Supramolecular Structures in Amorphous Butadiene-Styrene Block Copolymers," *J. Macromol. Sci., Phys.,* B6, no. 1 (1972), 167.

26. Samuels, S.L. and Wilkes, G.L., "Anisotropic Superstructure in Segmented Polyurethanes as Measured by Photographic Light Scattering," *J. Polym. Sci., Part B,* 9 (1971), 761.

27. Pimentel, G.C. and McClellan, A.L., *The Hydrogen Bond,* San Francisco: W.H. Freeman (1960), 70.

28. Estes, G.M., Seymour, R.W., and Cooper, S.L., "Infrared Studies of Segmented Polyurethane Elastomers. II. Infrared Dichroism," *Macromolecules,* 4 (1971),452.

29. Harrell, L.L., Jr., "Segmented Polyurethanes. Properties as a Function of Segmented Size and Distribution," *Macromolecules,* 2 (1969), 607.

30. Wolkenstein, M.W. and Ptitsyn, O.B., "The Transition into the Vitreous State and the Hydrogen Bonds," in *Hydrogen Bonding,* D. Hadzi, ed., New York: Pergamon Press (1959), 489.

31. Hannon, M.J. and Koenig, J.L., "Infrared Studies of Cryogenic Transitions in Poly (ethylene) Terephthalate," *J. Polym. Sci., Part A-2,* 7 (1969), 1085.

32. Fraser, R.D.B., "The Interpretation of Infrared Dichroism in Fibrous Protein Structures," *J. Chem. Phys.,* 21 (1953), 1511.
    Fraser, R.D.B., "Interpretation of Infrared Dichroism in Axially Oriented Polymers," *J. Chem. Phys.,* 28 (1958), 1113.

33. MacKnight, W.J. and Wang, M., "An Infrared Study of Hydrogen Bonding in Hard Polyurethanes," *J. Polym. Sci., Part C,* submitted for publication.

12. Rheo-Optical Studies of Block and Segmented Copolymers

GARTH L. WILKES and SAM L. SAMUELS

Polymer Materials Program
Department of Chemical Engineering
Princeton University
Princeton, New Jersey

ABSTRACT

A review is made of rheo-optical studies on block and segmented copolymers. Some detail is given on the principles of the basic rheo-optical techniques of birefringence, X-ray diffraction, dichroism, and small-angle light scattering. Numerous examples from the literature are included to illustrate the usefulness of the rheo-optical methods and, where possible, correlation of these data is made with the results obtained by more commonly known characterization methods such as microscopy.

Meaning of Rheo-Optics

The term "rheo-optics" is a relatively new word in the area of polymer science, but it is one that has assumed considerable meaning, particularly to the polymer physicist. The term itself originated in the laboratory of Professor R. S. Stein, the pioneer of rheo-optical techniques. The exact definition of this term is, however, not precise; rather, it is meant to encompass those techniques dependent upon the utilization of electromagnetic radiation (optics) in studying the deformation and flow (rheology) of polymers, with particular emphasis on polymeric solids. Furthermore, the term generally applies to techniques nondestructive in nature. In this paper the only methods considered under this heading are small-angle light scattering, dichroism, birefringence, X-ray diffraction, fluorescence, and Raman scattering. The characterization of deformation behavior by the latter two methods is still in its infancy, but it is anticipated that Raman spectroscopy in particular will be of future benefit. Although many other techniques could also be considered rheo-optical, such as microscopy and second-moment NMR studies, this has not been the case, probably because the term "rheo-optics" has been associated only with the main methods utilized by Professor Stein.

225

Staying within this "traditional" framework of the definition, we will attempt to review the *application* of these methods to systems of block and segmented copolymers. We have purposely included a considerable number of examples and figures from the literature in order to aid the reader with the type of data generated by rheo-optical studies. When possible, we have also attempted to confirm the usefulness of these methods by supporting the rheo-optical results with microscopy or mechanical results. In many cases, direct microscopy studies may not be possible, e.g., under dynamic conditions where only rheo-optical data can be obtained. We have also included a brief section on the principles and theory of the techniques. We hope in this way to familiarize and inspire those not acquainted with the aspects of rheo-optics.

General Background and Considerations of Structure–Property Relationships of Copolymer Systems

The advent of various synthetic random or nearly random copolymers was hardly a startling new development, considering the number of these systems routinely prepared by nature in all living systems (e.g., proteins)*. However, by making grafts or forming blocks or segments of two mutually incompatible homopolymer species, polymer chemists created materials that could assume a variety of unique morphologies and thereby a diverse range of properties. Some of the initial work illustrating this point was that of Merrett who grafted methyl methacrylate side chains onto a natural-rubber backbone [1]. His mechanical measurements on these systems with different processing histories led to the conclusion that some microphase separation was possible. Due to the graft structure, however, the degree of phase separation was somewhat restricted. This and other studies prompted the development of the linear block (A–B) systems, which are not configurationally or spatially restricted from undergoing microphase separation and, in fact, do display domain formation if incompatibility exists between the two block components. The work of Bradford and Vanzo [2] clearly confirmed this point. They prepared an A–B styrene-butadiene high-molecular-weight block system by a two-stage anionic polymerization scheme. Solution-cast films of these materials clearly showed the presence of microphase separation when investigated by transmission electron microscopy (TEM) using replication methods. Figure 1 gives one example of the observed behavior; it may be noted that the dimensions of the somewhat periodic spacing could be correlated with the molecular size of the block components. Although microphase separation was clearly documented in the A–B

---

*This is not to imply that protein structure is random, for it is synthesized by a specific recipe; relative to an alternating or block copolymer structure, however, it may be considered nearly random in many cases.

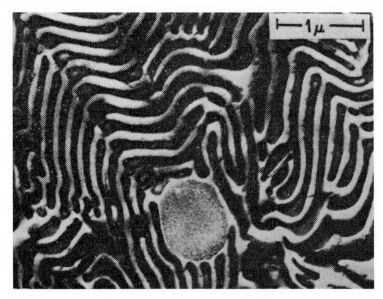

Figure 1. Transmission electron micrograph of a solution-cast film of an A–B block copolymer. (Courtesy of E. B. Bradford [2].)

systems, the mechanical properties of these systems were not well suited to promote their application as structural materials. Because of their "liquid-crystalline-like" characteristics, however, they have continued to be of interest for other reasons.

At the same time the A–B systems were under investigation, triblock (A–B–A or S–B–S) systems were also developed, one of the initial and foremost papers being by Hendus et al. [3]. This work illustrated the variation in mechanical behavior with component ratio and, furthermore, showed that these triblock systems of incompatible components were also prone to microphase separation, now termed domain formation. The evidence for this is commented on later.

Because the mechanical properties of these systems were appealing with respect to application, they prompted a wide number of studies by workers throughout the world. Such studies ranged from basic synthesis routes to tailor making block systems for particular properties based on knowledge of homopolymer behavior (e.g., glass transition, thermal stability, etc.). These latter studies have led to a wide variety of block systems such as those including the component pairs of styrene-isoprene, piperazine copolyamides, styrene-ethylene oxide (SEO), styrene-butadiene, ethylene-propylene (EP), sulfone-butadiene, and dimethylsiloxane-carbonate. Included in some of the above are systems displaying some crystallinity that may originate in only one of the two blocks (e.g., ethylene oxide in the SEO system) or in both components (as in the ethylene-propylene system). This aspect should be emphasized; viz., there are A-B-A systems that are completely amorphous and some that are partially crystalline.

Following the development of the block polymers, the segmented systems, sometimes called hard–soft segmented polymers, also appeared on a commercial level. These have been based primarily on urethane linkages between a soft segment such as tetramethylene oxide and an aromatic or nonaromatic isocyanate chain extended with a glycol (hard segment). These systems are similar to the block systems in that they generally show domain formation, but on a smaller scale, and they may have crystalline character if the "hard" segment crystallizes. In the latter case it has recently been shown by these authors that in certain cases these materials may develop large anisotropic spherulites, several microns in size [4].

During the development of these aforementioned polymers, there have been numerous studies to elucidate the structure–property relationships that determine behavior. The use of rheo-optical techniques have aided in many of these studies, and it is the purpose of this paper to review the results.

Principles of Rheo-Optical Methods

As previously mentioned, the number of methods included under the category of rheo-optics could be increased over the traditional sense of the term. However, we consider only techniques in the traditional sense, viz., birefringence, dichroism, X-ray diffraction, and small-angle light scattering. Some comments are made on the methods of polarized fluorescence and Raman scattering, but since these techniques have not been utilized to any great degree in studying the rheo-optical properties of the systems on which this paper is based, the discussion is very limited.

Birefringence

Birefringence, Δ, is defined as the difference between refractive indexes in two orthogonal directions (i, j) as measured with polarized light; i.e.,

$$\Delta = n_i - n_j. \tag{1}$$

A finite value of Δ implies, but does not require, optical anisotropy. Since the character of a polymer chain is such that there is generally a finite difference in the polarizability (refractive index) parallel to the backbone versus that perpendicular to it, the birefringence is a parameter that can be used to follow orientation. As the chains align, the magnitude of the difference between the refractive index parallel, $n_\parallel$, and perpendicular, $n_\perp$, to the orientation direction will increase. The sign of the birefringence will depend on the polarizability characteristics of the chain, i.e., the relative magnitudes of $n_\parallel$ and $n_\perp$. For example, polyethylene has a higher polarizability along the backbone ($n_\parallel > n_\perp$), while polystyrene has a greater polarizability perpendicular to the chain ($n_\perp > n_\parallel$) because of the highly polarizable phenyl ring.

Before proceeding to relate orientation and birefringence, it is necessary to point out that a finite birefringence may arise from other causes. It can arise due to bond stretching or distortion, which may occur during the deformation of glasses; in this case it is called deformation birefringence. Another important origin is form birefringence, which can arise when the system in question has at least two phases differing in refractive index, where the dimensions of the component phases are on the order of the wavelength of light. These considerations may become significantly important with several of the copolymer systems. Specifically, one has domain or microphase separation having regional dimensions of the above magnitude, and the refractive index difference between components is finite, e.g., polystyrene, $n = 1.56$, and polybutadiene, $n = 1.46$. The importance of birefringence is illustrated in the following section of this paper.

Birefringence gives only an *average* orientation index for the system, since it depends only on the second moment of the molecular orientation distribution. In single-component systems this is not a major hindrance, but in multicomponent systems a single birefringence value does not indicate how each component changes in orientation. Assuming additivity of birefringence, one can write

$$\Delta = \sum_i \phi_i \Delta_i + \sum_i \phi_i \Delta_{id} + \Delta_f, \tag{2}$$

where ϕ_i is the volume fraction of the ith component, Δ_i and Δ_{id} are the orientation and deformation birefringences, respectively, and Δ_f is the form birefringence. Since deformation birefringence will generally be considerably smaller than the orientation contribution, this term is neglected. Rewriting Equation (2) for a two-component (e.g., block copolymer) system in terms of orientation functions, we have

$$\Delta = \phi_A \Delta_A^o f_A + \phi_B \Delta_B^o f_B + \Delta_f, \tag{3}$$

where Δ^o is the intrinsic birefringence and represents the value of the birefringence for the perfectly oriented state; and

$$f \equiv \left(\frac{3\langle \cos^2 \theta_i \rangle - 1}{2} \right), \tag{4}$$

where $\langle \cos^2 \theta_i \rangle$ is the average value of the squared cosine of the angle that the chain axis of component i makes with the stretch axis. In Equation (4), f is commonly known as the Hermans' orientation function and is proportional to the second moment of the orientation distribution. Even if Δ_f is small, Equation (3) shows that measurement of Δ alone does not provide information about the orientation behavior of the *individual* components. To obtain this, one requires a knowledge of the orientation behavior (orientation function) of one of the components, which may be obtained by another method such as dichroism.

Birefringence and rubber elasticity. In the classic work of Kuhn and Guhn [5] regarding the statistical segment treatment of rubberlike homopolymers, the birefringence was related to the extension ratio λ by

$$\Delta = \frac{2}{45}\pi \frac{(\bar{n}^2 + 2)^2}{\bar{n}} N_c(b_1 - b_2)(\lambda^2 - 1/\lambda), \qquad (5)$$

where $\bar{n}$ is the average refractive index, N_c is the number of network chains per unit volume, and b_1 and b_2 represent the parallel and perpendicular polarizabilities of the statistical segment. When this is combined with the Gaussian elasticity equation for the stress σ (based on actual cross-sectional area), as a function of λ, one obtains the well-known stress optical law:

$$SOC = \frac{\Delta}{\sigma} = \frac{2\pi}{45kT} \frac{(\bar{n}^2 + 2)^2}{\bar{n}} (b_1 - b_2), \qquad (6)$$

where it is noted that the stress optical coefficient, SOC, is a function of only intrinsic material characteristics, assuming $(b_1 - b_2)$ is constant.* Equation (6) also has particular usefulness regarding the block, graft, and segmented rubbery systems, since a measurement of the SOC provides an index of the ideality of the rubberlike behavior. Furthermore, one can calculate a value of the anisotropy $(b_1 - b_2)$ from Equation (6), but this clearly has limited significance in a two-component or multicomponent system. Extensions of the above statistical segment approach and SOC considerations to two-component systems are discussed in the applications section of this paper.

The experimental methods of determining birefringence have been discussed in some detail elsewhere by one of us [6] and are not presented here. In general the measurement is simple and requires relatively little investment. Furthermore, one can make such measurements under static, dynamic (oscillatory), and/or high-speed deformation conditions [7,8]. Each method offers certain insight into the structure–property relationships.

Dichroism

This technique is based on the selective absorption of polarized radiation by the various chromophores in a polymer chain. Depending upon the absorbing

---

*The term $(b_1 - b_2)$ has been shown to display some small temperature dependence.

frequency of the chromophore and the nature of the absorption mode, the measurements may involve ultraviolet, visible, or infrared radiation.

Associated with a given absorption mode is a transition moment, μ, which is directed along the principal axis of the absorption ellipsoid, e.g., parallel to the double bond of a carbonyl group for a carbonyl stretch absorption. With polarized radiation, the magnitude of the absorption, A, of a given chromophore will be proportional to the square of the dot product between the transition moment vector, μ, and the vector, $\mathbf{P}$, representing the polarized ray:

$$A \sim (\mu \cdot \mathbf{P})^2 = |\mu|^2 |\mathbf{P}|^2 \cos^2 \theta. \tag{7}$$

The average orientation, $\langle \cos^2 \theta \rangle$, of a chromophoric group can therefore be determined by using polarized radiation with the proper absorption frequency. Since one generally desires the orientation of the chain backbone, the angle, α, between the vector, μ, and the chain axis must be known. Based on the above considerations, it can be shown [9] that by measuring the desired absorption with the polarizer aligned parallel, $A_\parallel$, and perpendicular, $A_\perp$, to the deformation axis, the chain orientation can be determined by

$$f_i = \left(\frac{D_o + 2}{D_o - 1}\right)\left(\frac{D - 1}{D + 2}\right), \tag{8}$$

where

$$D_o = 2 \cot^2 \alpha, \tag{9}$$

and

$$D \equiv A_\parallel / A_\perp, \tag{10}$$

where D is, by definition, the dichroism. The orientation function, f, in Equation (8) has been given a subscript i to denote that this average orientation is to be associated with the region of the chain where the chosen chromophoric group is located. This point has particular ramifications in multicomponent or multiphase systems such as the copolymers under discussion. For example, Figure 2 shows a hypothetical triblock copolymer molecule that has undergone deformation. Although the model is oversimplified, it shows that the average orientation is different in the two blocks. If, then, a *different* chromophoric unit of known μ can be found in each of the two different blocks, by measuring D one could obtain the individual values of f for *each* component. Therefore, dichroism offers a particular advantage over birefringence because, in principle, individual-component orientation behavior can be determined.

Although the measurement of dichroism is straightforward, it suffers from the requirement that only thin films be used as the sample material, the optimal thickness being dependent upon the extinction coefficient of the chromophoric group to be investigated. The exception to this case occurs if the method of

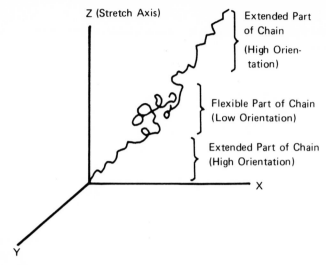

Figure 2. Illustration showing how a block copolymer could display different orientations for its individual components.

attenuated total reflectance (ATR) is used with polarized radiation. This method measures only the local surface orientation, however, and the data are somewhat more difficult to interpret due to the effect of reflectance on the polarization.

X-Ray Diffraction

Because of the number of complete treatises on wide-angle diffraction, there is little need to review the basic principles of this method [10-12]. However, some general comments are in order with respect to its application to the measurement of the degree and time dependence of crystalline orientation.

Figure 3 illustrates a typical diffraction experiment to determine the orientation in a crystalline material where one wishes to specify the spatial orientation of the crystals (unit cell) relative to a fixed coordinate system based on the sample axes. The general procedure is to measure the orientation of the normal, ρ_{hkl}, to a given set of (hkl) planes that relate to the crystallographic axis in question; for example, the normal of an (001) plane in polyethylene is parallel to the c axis as well as the chain axis. In Figure 3, the sample axes are denoted as Z_s, Y_s, and X_s, and the laboratory coordinates as X, Y, and Z. From this geometrical arrangement, it follows that the average orientation, $\langle \cos^2 \theta \rangle_{\rho_{hkl}, Z_s}$,

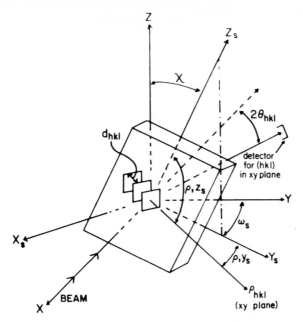

Figure 3. Schematic of the general geometrical characteristics of an X-ray diffraction experiment. Sample axes are X_S, Y_S, and Z_S, and the fixed laboratory coordinates are given by X, Y, and Z.

can be determined from

$$\langle \cos^2 \theta \rangle_{\rho_{hkl}, Z_S} = \frac{\int_0^{2\pi} \int_{-\pi/2}^{\pi/2} I(\pi/2 - \chi, \theta - \omega) \cos^2(\pi/2 - \chi) \sin(\pi/2 - \chi) d\chi d\omega}{\int_0^{2\pi} \int_{-\pi/2}^{\pi/2} I(\pi/2 - \chi, \theta - \omega) \sin(\pi/2 - \chi) d\chi d\omega},$$

(11)

where $I(\pi/2 - \chi, \theta - \omega)$ is the measured intensity, and the subscripts of *hkl* indicate the planes to which the orientation refers. If cylindrical symmetry exists about Z_S, then integration over ω in Equation (11) is not required. It is important to point out that by knowing the angular dependence of the diffracted intensity, one can obtain any moment of the orientation distribution and, hence, its *complete spatial* description, which may be presented in the form of a pole-figure diagram as shown in Figure 4. This diagram is simply a stereographic contour plot of the orientation of a specific (*hkl*) normal. Since higher moments

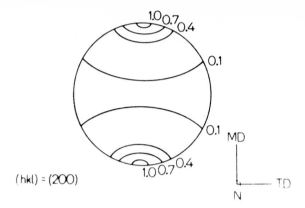

Figure 4. Hypothetical pole figure for a (200) plane. Since the normal to a (200) plane is parallel to the A axis, this plot directly reflects the fact that the A axis is oriented along the machine direction (MD). Numbers on contour lines represent relative intensities.

of the orientation distribution can be calculated by Equation (11), it can be seen that diffraction allows a much more complete description of orientation than do the methods of birefringence and/or dichroism (second-moment methods). Although the diffraction technique has been applied to elucidate amorphous orientation [13], its greater degree of success is limited to, and is best suited for, semicrystalline systems. In the latter cases (e.g., ethylene oxide-styrene block copolymers), this method can be applied in conjunction with the other rheo-optical methods to aid in the characterization of the orientation behavior of all components.

It has recently been shown by Stein and Oda [14] that rapid changes (~ milliseconds) in crystalline orientation can be followed by the diffraction method. This new approach may well be superior to the former dynamic X-ray method also developed by Stein [7] and may be useful in studying the time-dependent orientation behavior of semicrystalline block or segmented systems.

Small-Angle Diffraction

The method of small-angle diffraction is typically utilized in one of two ways. One is to compare the absolute intensity of the scattered radiation with those intensities calculated on the basis of various model systems. Since this paper is directed at the block and segmented copolymers prone to domain formation, a brief consideration of the scattering from a two-phase system is in order. If one assumes a sharp boundary between the two components (matrix, A, and particle, B), then one can write [10]

$$(\rho - \bar{\rho})^2 = (\rho_A - \rho_B)\,\omega_A\,\omega_B, \tag{12}$$

where the left side of Equation (12) is called the scattering power, or mean-squared fluctuation in electron density, and the terms on the right are the respective electron densities (ρ), their average ($\bar{\rho}$), and volume fractions (ω) of components A and B. The scattering power is, in turn, related to the angular scattered intensity:

$$(\rho - \bar{\rho})^2 = \frac{4\pi\bar{\rho}Q}{\gamma\eta}, \tag{13}$$

where η is the thickness of the sample, and γ is equal to $7.9 \times 10^{-26} \lambda^2$, λ being the wavelength of X-ray radiation. The term Q is known as the invariant and is defined as

$$Q = \int_0^\infty I(m)\, m^2\, dm, \tag{14}$$

where $I(m)$ is the angle-dependent absolute intensity, and m is equal to $2a \sin(\phi/2)$, where a is the detector-to-sample distance, and ϕ is the radial scattering angle. It follows that the existence of a two-phase system can then be verified by this approach.

The second use of small-angle scattering is more qualitative and concerns the interpretation of distinct intensity maxima as noted by photographic techniques. Here again, a model is generally proposed, and the Bragg spacing determined from the film is correlated with such factors as interparticle or interlamellar distances.

Small-Angle Light Scattering

Light scattering has proven to be a useful tool in understanding the internal superstructure of polymeric materials. Fluctuations in both optical anisotropy and density are of particular importance, although other information can be obtained. The method has the advantage that it is nondestructive and can provide information regarding structure having dimensions on the order of the wavelength of light. Thus, information is obtained that cannot always be observed by optical microscopy—this is particularly so when the experiment is dynamic in nature (e.g., stretching).

There are two common experimental approaches: the quantitative measurement of intensity (absolute or relative) by photometric methods, and the so-called photographic technique, which involves taking a photograph of the light-scattering pattern at small angles (see Figure 5). The former method is best suited for detailed quantitative information such as the mean-square fluctuation in density, $\langle\eta^2\rangle$, the evaluation of persistance length, and for comparison with

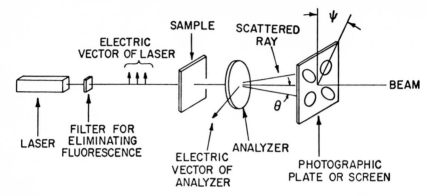

Figure 5. Typical optical setup used for the photographic light-scattering experiment. The analyzer may be crossed (H_v) or aligned parallel (V_v) to the polarizer, each orientation providing different information.

theoretical calculations. The photographic method suffers somewhat by not providing quantitative intensities, but it is particularly useful for rapid and qualitative studies, particularly where optical anisotropy is present. For example, the photographic method is particularly well suited for providing information on spherulite size, type of molecular packing, and change in superstructure during or following deformation. Figure 6a shows H_v and V_v patterns* typical of undeformed polyethylene, while Figure 6b shows the H_v and V_v patterns for a deformed polyethylene sample. Upon deformation, the patterns have changed, and this change can be directly related to the deformation of the anisotropic spherulites. An example of scattering from a different type of superstructure is illustrated in Figures 7a and 7b, which show H_v and V_v patterns for an undeformed and a deformed film of a polypeptide, poly(glutamic acid). These patterns differ from those of polyethylene and can be interpreted in terms of anisotropic rods rather than spherulites. Clearly, then, the scattering pattern gives a rapid indication of the morphology and presence of (or lack of) optical anisotropy. Details on experimental considerations have been summarized elsewhere [16].

Two approaches are used in conjunction with the interpretation of scattering results: the statistical approach involves correlation functions relating fluctuations over spatial regions; the distinct geometrical model approach involves calculating the scattering from a model (e.g., spherulite) and comparing it with

*H_v refers to a horizontal (H) and vertical (v) alignment of the electric vectors of the analyzer and polarizer, respectively. If a stretch axis is present, it would be aligned along the v direction. Consequently, in a V_v orientation both electric vectors are aligned in a vertical direction. The stretch axis would also be parallel to these vectors. In an H_h position, the stretch axis would be perpendicular to both analyzer and polarizer vectors.

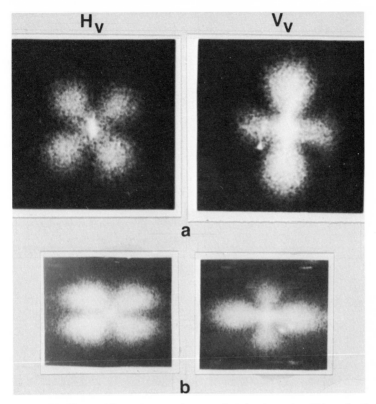

Figure 6. SALS H_v and V_v patterns taken on low-density spherulitic polyethylene: (a) undeformed; (b) stretched 20% (stretch axis is vertical).

the observed patterns. In practice, both are necessary, but one may be more suitable than the other for specific purposes, as is illustrated in the section on applications. In general the model approach is favored since one would prefer, if possible, to relate the scattering to a distinct structural entity such as a rod or sphere. One can calculate the scattering from such systems (isotropic or anisotropic) by summing the amplitudes of the scattered intensity over all volume elements of the unit in question. Where such structural entities do not exist, correlation functions (statistical approach) are employed and are comparable in nature to those used in X-ray scattering from liquids. Specifically, the scattering is correlated to the statistical fluctuations in either density (isotropic scattering) or orientation of anisotropic units (anisotropic scattering). Even when the scattering can be qualitatively explained with the model approach, there is generally a component that can only be accounted for with the statistical approach.

From this discussion it is evident that light scattering provides a type of in-

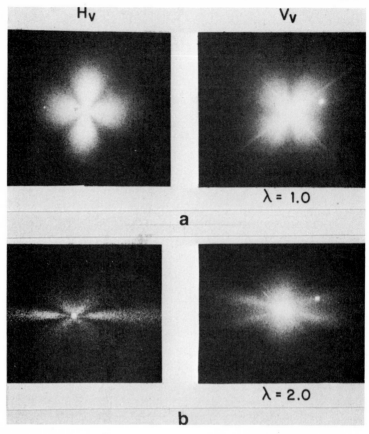

Figure 7. SALS H_v and V_v patterns taken on a film of poly(glutamic acid) cast from DMF: (a) undeformed; (b) stretched about 100% (stretch axis is vertical).

formation different from that given by the methods of birefringence, dichroism, and wide-angle diffraction. In fact, the latter methods are usually used in conjunction with quantitative orientation measurements, while the former (as well as small-angle X-ray scattering) is used for obtaining information concerning homogeneity and deformation of structure.

Polarized Fluorescence and Polarized Raman Scattering

Because these methods have not been applied directly to the copolymer systems under consideration, none of the details of these techniques is presented here. It is worthwhile to point out that both methods can, in principle, lead to both the second and fourth moments of the orientation function of the *individual* components. Results from these techniques have not been obtained because some of the important assumptions necessary to interpret the data have not been verified. Both methods are relatively straightforward experimentally and are nondestructive under proper conditions. The interested reader should consult the appropriate references for further information on the fluorescence [17,18] and Raman methods [19].

Applications of Rheo-Optics

To illustrate how rheo-optical techniques can aid in structure-property investigations of block or segmented copolymers, the discussion, although focused at times on a single technique, is correlated, when possible, with data obtained by other rheo-optical or related methods. Although this results in some overlap in presentation, it nevertheless proves the most fruitful approach.

Small-Angle X-Ray Diffraction (SAX)

This method, in conjunction with transmission electron microscopy (TEM), confirmed the hypothesis of microphase separation, i.e., domain formation, in the A–B, A–B–A, and segmented polymers. The first SAX evidence (film technique) was presented by Hendus et al. [3] and is given in Figure 8. These data are from an S–B–S system in the undeformed and deformed states. Distinct maxima can be noted, and the Bragg spacing associated with them is of the order of 350 Å, which is typical and relates to the domain structure directly observed in the TEM when heavy-metal staining is used. An example of an undeformed and a deformed S–B–S system is shown in Figure 9 [20]. The intensity maximum is a consequence of the well-developed periodicity in electron density, i.e., the difference associated with the styrene and the butadiene phases. In the deformed samples, the equatorial and meridian spacings can be correlated with the

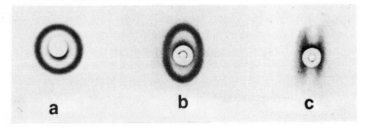

Figure 8. Small-angle X-ray patterns taken on a solution-cast S–B–S block copolymer: (a) undeformed; (b) drawn beyond necking and then released; (c) redrawn to 100% strain. For (b) and (c) the stretch axis is horizontal. (Courtesy of H. Hendus [3].)

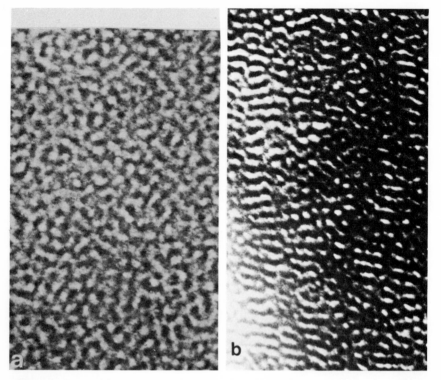

Figure 9. Transmission electron micrographs of an S–B–S (Kraton 101) block copolymer: (a) undeformed; (b) drawn. (Courtesy of S. L. Aggarwal [20].)

dimensional anisotropy of the domains. The fact that only a single maxima is found implies that there are local irregularities in the periodicities that destructively interfere with diffraction from long-range order (over several domains). By careful film preparation, however, this order can be increased and more well-defined SAX patterns can be obtained. One of the most striking examples of this has been given by Keller et al. [21,22] for the well-studied Kraton 102 S–B–S system. Figure 10a shows a SAX pattern taken parallel to the axis of an extruded plug of this material, while Figures 10b and 10c show the corresponding TEM micrographs of the surface perpendicular and parallel to the X-ray beam. The regularity and long-range periodicity in structure explain the occurrence of the "macroscopic single-crystal" X-ray pattern. This work is discussed further in the next section.

An example of the use of the SAX film technique involving segmented urethane systems was presented by Samuels and Wilkes [26], who used SAX to confirm the existence of small domains (<100 Å). Although these systems were semicrystalline, there were strong reasons to associate this SAX spacing with domain structure rather than the usual lamellar long spacing.

Absolute SAX intensities have been measured for both the A–B–A block [23] and segmented urethane systems [24]. The earlier study by Legrand utilized Equation (12) based on a two-phase model. Knowing the volume fractions and respective electron densities of the two block components, he could then compare the experimentally observed intensities with the calculated value based on the mean-square fluctuation in electron density $\langle \rho^2 \rangle$ (sharp boundaries between phases assumed). His overall results for several types of block copolymers indicated that $\langle \rho^2 \rangle_{exp} < \langle \rho^2 \rangle_{theory}$, thereby giving further support to the belief that there is a gradient in electron density near the boundary region that is due to the partial mixing of the two components. This gradient would explain the lower value of $\langle \rho^2 \rangle$ and is in line with the theory of Williams [27] and the experimental results of Kaelble [25], Beecher [20], and Samuels and Wilkes [26]. It would be interesting to perform a similar SAX study where the solubility parameter of the casting solvent or processing history varied for a group of A–B–A systems. Since the type of casting medium influences phase separation [20], the SAX technique might help elucidate the character of the boundary and thereby the degree of mixing of the components.

A similar two-phase treatment has been applied using the photometric method of small-angle light scattering (SALS) for both A–B [30] and A–B–A [28,30] systems. This treatment is based on the Debye–Bueche [31] theory, which predicts that for an isotropic two-phase system the scattered intensity, $I(h)$, can be written as

$$I(h) = CV\beta^2 \phi_A (1 - \phi_A) \int_V \gamma(r) \exp\left[i(h \cdot r)\right] dr, \qquad (15)$$

where C is a constant, V is the scattering volume, ϕ_A is the volume fraction of one of the components, β^2 is the polarizability difference between the two components, and $h = ks$, where $k = 2\pi/\lambda'$ and $|s| = 2\sin(\theta/2)$, λ' being the wavelength of light in the medium, and θ the scattering angle.

The quantity $\gamma(r)$ is known as the correlation function and can be expressed as

$$\gamma(r) = \frac{1}{\bar{\eta}^2} \int_V \eta(x)\,\eta(x + r)\,dx, \qquad (16)$$

where $\eta(x)$ and $\eta(x + r)$ are the fluctuations in polarizability from the mean of the system at point x and $x + r$. It can also be shown that $\bar{\eta}^2$ is equal to $\beta^2 \phi_A (1 - \phi_A)$ for a two-phase system. By a Fourier transform of the scattered intensity $I(h)$, one can obtain the parameters $\bar{\eta}^2$ and the correlation function $\gamma(r)$. From the same data one can also extract information regarding the degree of specific surface or boundary area between the two phases. The theory of Debye and Bueche has been extended to anisotropic systems by Goldstein and Michalek [32] and Stein and Wilson [33] and has been applied to several

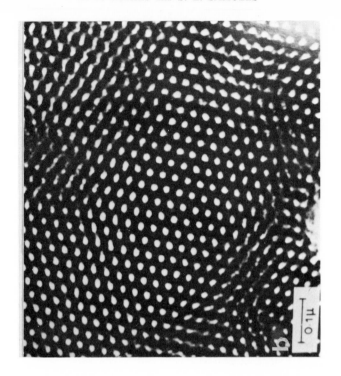

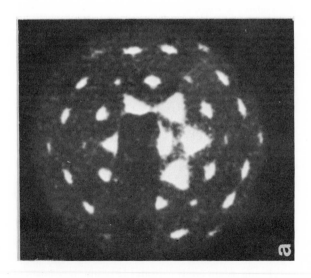

Figure 10. (a) Small-angle X-ray pattern for an extruded cylinder of an S–B–S (Kraton 102) block copolymer. X-ray beam was directed parallel to the extrusion axis. Corresponding transmission electron micrograph on (b) perpendicular and (c) parallel film sections of the extruded cylinder above. (Courtesy of A. Keller [21,22].)

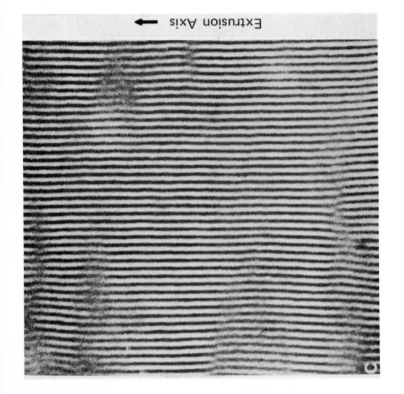

Extrusion Axis

polymeric systems [30,34–36]. Using various A–B block copolymers (some mixed with the respective homopolymers), Kawai et al. [30] tested the basic theory above. These polymers provide a suitable model system since one can control the shape and quantity of the second phase. Furthermore, since the polarizability differences are known, comparison between theory and experiment can be made. By using TEM studies in conjunction with the scattering investigations, a better understanding of the correlation function $\bar{\eta}^2$ and specific surface values would be gained. In brief, the study of Kawai et al. gave considerable support to the merits of this theory. They also showed that the average size of the dispersed phase determined by the scattering method (specific surface data) compared well with the TEM results (see Figure 11).

Kawai et al. [29] have also used photometric SALS in deformation studies of A–B and A–B–A systems. These investigations were made in conjunction with intensity measurements of small-angle X-ray scattering. The X-ray results suggested that local domain regions do not deform affinely, the degree of deviation being dependent upon the extension ratio. The SALS also supported this conclusion.

Photometric SALS has also been utilized by Clough and Schneider [37] in

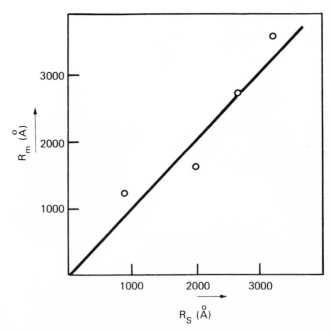

Figure 11. Plot showing comparison of the radius, R, of the "domain" phase as deduced from scattering measurements, R_s, with the radius observed by TEM investigations, R_M. (Courtesy of H. Kawai [30].)

their structural investigations on segmented urethane elastomers. These studies, carried out in conjunction with small-angle X-ray intensity measurements, gave support for domain formation in these systems, but on a smaller-size scale than that noted for the usual block copolymers.

Birefringence

Probably the greatest use of birefringence with respect to the characterization of block copolymers is in the stress optical law previously given in Equation 6. This relationship is used to test the degree of ideal rubber behavior (Gaussian case) by the measurement of the SOC. Figure 12 shows the results of such investigations by Wilkes and Stein [38] made under stress-relaxation conditions at different temperatures on an S–B–S system cast from two different solvents, toluene and methyl ethyl ketone (MEK). Figure 12a shows the SOC as a function of time for different temperatures for the toluene-cast film, while Figure 12b shows that for the MEK-cast film. The point of interest is that the SOC is nearly constant for the toluene material, while that for the MEK material increases with increasing time and temperature. Thus, the toluene film closely follows Gaussian predictions, while the MEK film does not. Similar conclusions could be reached from stress–strain measurements, since the toluene-cast film displays a rubberlike behavior, while the MEK-cast material displays a distinct yield point (and may even neck), as illustrated by the stress–strain data of Beecher et al. [20] for the same material (Figures 13a and 13b). In these latter data, the effect of mechanical hysteresis is noted, and there is a considerable difference in the behavior of the MEK- and toluene-cast films. Figures 14a and 14b show the optical hysteresis as a function of strain for toluene-cast (or compression-molded) and MEK-cast films. One notes a greater hysteresis for the MEK material, as well as considerably lower birefringence values.

The above optical and mechanical data are interpreted via a two-phase system where the degree of phase separation is higher in the toluene-cast film (or compression-molded material) than in the MEK-cast material, this being in agreement with TEM studies [20]. That is, in the toluene films, the butadiene acts as the continuous-phase and load-bearing component, while the polystyrene domains act as a filler and macro-cross-linking points for the butadiene chains. For the MEK material there is more of a mixing of the PS and PB components, and thus a higher portion of the PS acts as a load-bearing component. This results not only in a higher initial modulus but also in a subsequent yielding upon deformation at room temperature, because the PS regions (all assumed to be in the glassy state) will break apart since this material is well below its glass-transition temperature ($T_g = 100°C$).

This explanation is supported by the birefringence data in that the values are higher for the toluene material, are less time-dependent (data not given), and

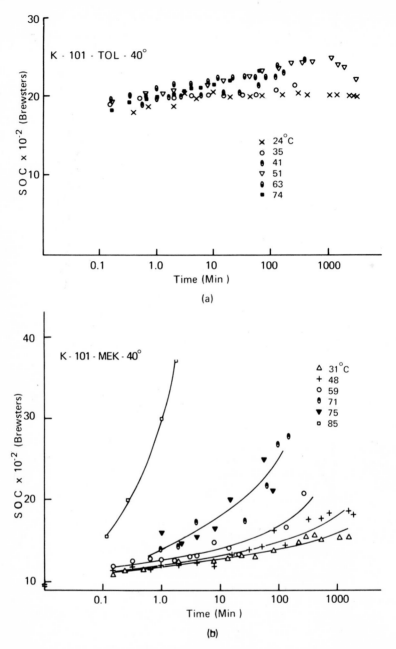

Figure 12. Plots of the stress optical coefficient vs. time for several temperatures during stress relaxation at a fixed strain of 100%: (a) film cast from toluene; (b) film cast from MEK. (Courtesy of Wiley & Sons from [38].)

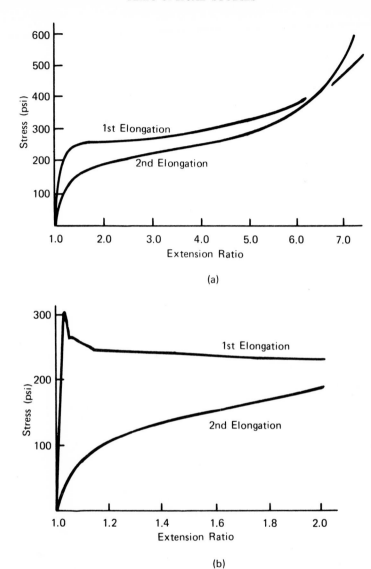

(a)

(b)

Figure 13. Stress–strain data for an S–B–S (Kraton 101) block copolymer cast from different solvents: (a) film cast from toluene; (b) film cast from MEK. (Courtesy of S. L. Aggarwal [20].)

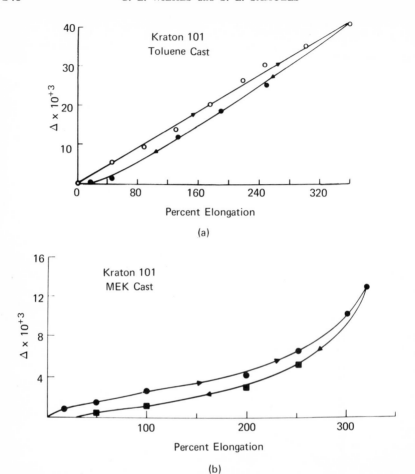

Figure 14. Plots showing birefringence vs. percent elongation for an S–B–S (Kraton 101) block copolymer. Arrows show the direction in which the data were obtained: (a) film cast from toluene (or compression-molded); (b) film cast from MEK.

show less hysteresis than do those of the MEK films. The higher birefringence values follow from the fact that the SOC of PB is about +3,200 Brewsters [39] as compared to -5,200 Brewsters for [40] PS (rubbery state) or +10 (glassy state) [41]. Thus, if the load-bearing component is mostly PB, the SOC and birefringence should be higher than if PS participates in the process. If this is true, one has to account for the fact that on the second elongation of the MEK material (following breakup of styrene domains), the birefringence is lower than the initial values and yet the stress–strain curve is rubberlike. This behavior can be rationalized on the basis that there is a much greater and faster relaxation in the MEK-cast than in the toluene-cast system, as can be noted from the con-

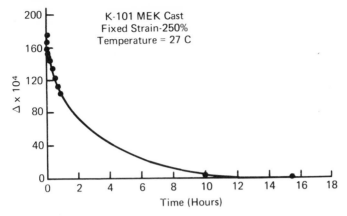

Figure 15. Plot of birefringence vs. time for a film of an S–B–S (Kraton 101) block co-polymer cast from MEK at 25°C. The strain was fixed at 250%.

siderable permanent set and rapid decay of birefringence at a fixed strain (see Figure 15). Similar behavior is not observed for the toluene (phase-separated) material, and, in fact, considerable amorphous orientation can be noted after long times, as determined by wide-angle X-ray diffraction for highly stretched films (see Figure 16). A further illustration of the difference in time-dependent behavior is given in Figure 17, which shows that a much greater difference exists in the magnitude of birefringence for the MEK material when measured at two very different strain rates. From these data it follows that some rapid relaxation

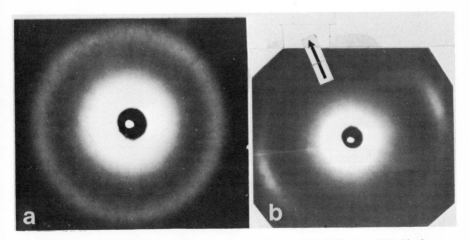

Figure 16. Wide-angle X-ray diffraction patterns for an S–B–S (Kraton 101) block co-polymer cast from toluene: (a) undeformed; (b) deformed to 4000%. Arrow represents the stretch axis.

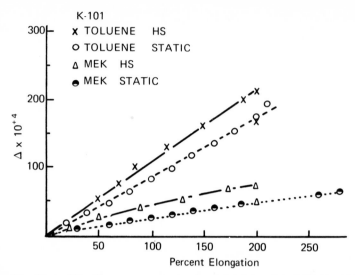

Figure 17. Plots of birefringence vs. percent elongation for an S–B–S (Kraton 101) block copolymer. The data measured at high speed (HS) conditions were obtained in less than 40 msec. The data recorded under static conditions took on the order of minutes. (Courtesy of Wiley & Sons from [38].)

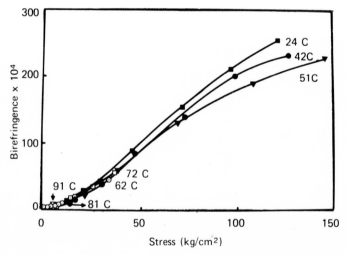

Figure 18. Plot of birefringence vs. stress for a compression-molded film of an S–B–S block copolymer (40 wt-% styrene) measured at several temperatures. (Courtesy of E. Fischer [42].)

also occurs in the toluene phase-separated materials, but the percentage is much lower.

Fischer and Henderson [42] have also investigated the birefringence behavior of compression-molded S–B–S systems.* One of the noteworthy results of these studies follows from Figure 18, where the birefringence is plotted as a function of the stress. One notes that for the lower temperatures there are three regions, each with a different slope. The first region of low stress is associated with partial deformation and yielding of PS, which makes up a part of the deforming matrix (hence, low birefringence and lower SOC). Further increase of stress results in an increase of birefringence, which is associated with PB orientation. Finally, higher stress (high elongation) leads to a lower dependence of birefringence on stress and is attributed to the onset of plastic deformation of PS, leading ultimately to fracture. Another point of interest in this study is illustrated in Figure 19 (showing how the SOC varies with temperature and styrene content) and is in line with the above discussion; i.e., the lower SOC is expected for the higher-styrene-content samples since the SOC of PS is lower than that of PB.

Using the data of Fischer and Henderson [42] for the toluene-cast films, one can plot the values of SOC against styrene content as in Figure 20. This shows that nearly a linear relationship exists over this region. It should be noted, how-

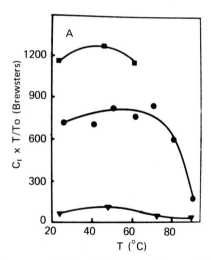

Figure 19. Plot of the stress optical coefficient versus temperature for three S–B–S block copolymers having different weight-percentages of styrene: ■ = 31% styrene; ● = 40%; ▼ = 49%. (Courtesy of E. Fischer [42].)

---

*Compression-molded films display properties very similar to those of toluene-cast films [28].

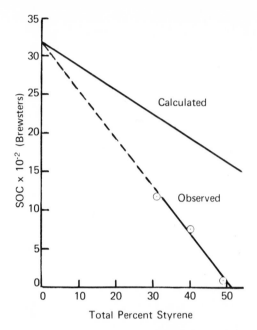

Figure 20. Plots of the stress optical coefficient vs. weight-percent styrene. The calculated plot is based on Equation (17) for the two cases described in the text. The data points were taken from Fischer and Henderson [42].

ever, that this relationship is based on a particular film history that gives rise to a fairly well defined two-phase system, and this behavior might not be observed in, for example, the MEK material.

If we assume additivity of birefringences and, furthermore, neglect form birefringence, which is reasonable for small deformations, then we can write the stress optical law as

$$SOC = \frac{\Delta}{\sigma} = \kappa_a \phi_a \frac{\Delta_a}{\sigma} + \kappa_b \phi_b \frac{\Delta_b}{\sigma}, \qquad (17)$$

where the terms κ_a and κ_b are related to the fraction of total stress that is borne by the respective components. If the stress field is homogeneous, for example, $(\kappa_a = \kappa_b = 1)$, Equation (17) becomes

$$SOC = \phi_a SOC_a + \phi_b SOC_b. \qquad (18)$$

On the other hand, if component a represents PB and if the load-bearing component is only PB, then $\kappa_a = 1$, $\kappa_b = 0$, and the experimental SOC is a function only of $\phi_a SOC_a$. In Figure 20 we have also plotted the theoretically calculated relationship for these two cases of stress distribution. Only one line is shown because the results are essentially identical and therefore illustrate the insensitivity of Equation (17) to different stress fields. In both cases we have assumed that the stress optical term correlated with PS is that of the glassy state, i.e., a SOC of +10 Brewsters.

From Figure 20 three observations can be made: (1) both theoretical cases exceed the observed behavior, particularly as the PS content increases; (2) the experimental line can be extrapolated to pure PB giving the SOC of PB; and (3) although there is little difference in the two theoretical plots, the stress fields are considerably different because the SOC of glassy PS is so small relative to that of the PB. Since it is difficult to conceive that the PS is the major load-bearing component, one is tempted to try a three-term SOC equation for a homogeneous stress field where the additional term is that associated with the amount of S in the rubbery phase (R), which might be expected at the interface region. T̲us, Equation (17) becomes

$$SOC = \phi_B\,SOC_B + \phi_{Sg}\,SOC_{Sg} + (1 - \phi_B - \phi_{Sg})\,SOC_{Sr}, \qquad (19)$$

where r and g refer to rubbery and glassy phases, and B and S refer to poly-butadiene and polystyrene. For brevity we carried out this calculation by setting the left side of Equation (19) equal to the experimentally measured values of Fischer and Henderson. Knowing ϕ_B, we could calculate the respective glassy and rubbery fractions of PS based on this relationship. The results are plotted in Figure 21, where two significant points are noted: (1) there is, somewhat

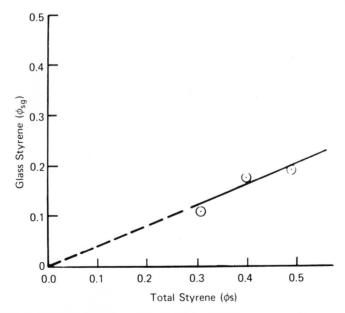

Figure 21. Plot of the volume fraction of glassy styrene, ϕ_{Sg}, vs. total percent styrene as calculated by Equation (19) and the data of Fischer and Henderson [42].

surprisingly, a nearly linear dependence of ϕ_{Sg} with total PS content (ϕ_S), which extrapolates to zero for $\phi_S = 0$, as would be expected; that is, as the total styrene content decreases, so does the glassy fraction; and (2) the fraction in the glassy state is less than that in the rubbery state, which seems surprising. This calculation has assumed a homogeneous stress field (series addition of components), which undoubtedly is an oversimplification. However, the results are rewarding and suggest further work in this area, since there has been considerable interest in the interfacial region of the PS–PB domains. These efforts could also be correlated with further model calculations of mechanical properties based on the approach of Takayanagi [43].

Thus far we have neglected the form contribution to the birefringence. As discussed earlier, this contribution could become finite for these two-phase systems of differing refractive indexes. This is true if there is a geometrical anisotropy to the dimensions of the phases, i.e., rodlike or platelike versus spherical. The classical theory of Wiener [44] treats various geometries and develops the equations for calculating the form contributions. A second reference that reviews many of the aspects of form birefringence is also suggested [45]. To briefly review one case, we will consider Figure 22, which shows long, thin rods of refractive index n_R dispersed in a matrix material of

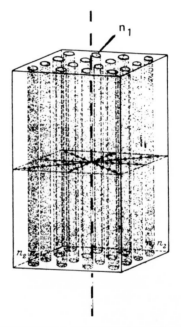

Figure 22. Illustration showing long, thin cylindrical rods of refractive index $n_1(n_R)$ embedded in a matrix of refractive index $n_2(n_M)$. (Taken from [45].)

refractive index n_M. If the separation distance between rods is small relative to the wavelength of light, then the form birefringence, Δ_f (reference axis is the rod axis), is given by [45]

$$\Delta_f = \frac{\phi_R \, \phi_M \, (n_R^2 - n_M^2)^2}{(n_A + n_B)\,[(\phi_R + 1)\, n_M^2 + \phi_M \, n_R^2]} \, , \tag{20}$$

where

$$n_A^2 = n_M^2 \frac{(\phi_R + 1)\, n_R^2 + \phi_M \, n_M^2}{(\phi_R + 1)\, n_M^2 + \phi_M \, n_R^2} \quad \text{and} \quad n_B^2 = \phi_R \, n_R^2 + \phi_M \, n_M^2.$$

Equation (20) shows that as $n_R \longrightarrow n_M$, Δ_f goes to zero, as expected. Furthermore, the sign of Δ_f will always be positive regardless of which refractive index is larger.

One might think this geometry is too idealized for the system under discussion, but recalling the TEM data in Figure 9b, we can see that in this case some characteristics of this rodlike geometry are apparent. Another more striking example of this has been presented by Folkes and Keller [21] in extruded and annealed plugs of Kraton S–B–S polymers. In these materials the geometrical arrangement of the PB and PS phases are in excellent agreement with the rodlike geometry (compare Figure 10c with Figure 22).

A section of this material was found by these same authors to give a positive birefringence (with respect to the direction parallel to the rod axes) of 5.15×10^{-4}. Since the volume fraction and refractive indexes of PS and PB were known, Equation (20) was utilized, and the calculated form anisotropy was found to be $4.92 \pm 1 \times 10^{-4}$, which is in excellent agreement with that observed. Although this might appear to be an extreme case of ideal form geometry, it should be recalled that these plugs were simply extruded, as might occur during a processing operation. Other well-documented examples of form birefringence in block or segmented polymers are lacking, but it is reasonable to expect that, particularly for highly deformed or extruded materials, the effects of form anistropy may influence optical measurements.

Dichroism

Compared to the method of birefringence, the dichroism technique of measuring orientation has been utilized relatively little in studying the orientation of block or segmented copolymers.

This is especially surprising since dichroism has the potential of being able to separate the component orientation, whereas birefringence can only provide an average of the orientation of all components. One study that indicates the advantage of the dichroism method over that of birefringence has been made by

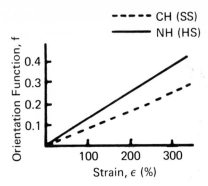

Figure 23. Plot of the Hermans' orientation function vs. strain for a polyether-urethane elastomer. NH and CH refer to the orientation of the hard segment (HS) and soft segment (SS), respectively. (Based on the data of Estes et al. [47].)

Cooper et al. in their studies of segmented polyurethane elastomers based on either polyester or polyether soft segments [46].

By selecting the chromophoric groups $-CH_2$ and $-NH$, which were assumed to be associated with the soft and hard segments, respectively, they illustrated the difference in the orientation behavior of the two types of segments. Figure 23 shows a plot of the Hermans' orientation function versus percent strain for the two types of segments in a polyether system. It is noted that considerably higher orientation is induced for the hard segment (stiffer segment). This behavior is also in line with the theoretical predictions of Shindo and Stein [47], Flory and Abe [48], and Nagai [49] for chains composed of different "statistical segment" types.

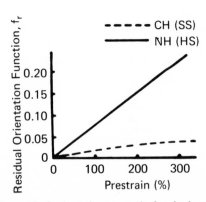

Figure 24. Plot of the *residual* orientation vs. strain for the hard segment (HS) and soft segment (SS) of a polyether-urethane elastomer that had been initially stretched to the given strain. (Based on the data of Estes et al. [47].)

Upon release of strain (prestrain) for these elastomers, the orientation did not totally disappear for either segment, as shown in Figure 24, and considerably more orientation is retained by the hard segment. Similar hard-segment behavior has also been noted in the wide-angle diffraction studies of Bonart [50] and has been attributed to the restrictions imposed by the formation of new hydrogen-bond bridges caused by the initial deformation. The higher retractibility of the soft segments is correlated with their degree of rubberlike behavior and lower involvement with hydrogen bonding. A further interesting point of this study is illustrated by Figure 25, where one notes the effect of an initial prestrain of 200 percent on the measured orientation during the second deformation. Specifically, the soft-segment orientation is again reversible, while that of the hard segment is not. Similar studies by Samuels and Wilkes and by Cooper [51] are underway for the series of polyurethane polymers discussed in the next section. These results should be of interest when compared to those just discussed, since in this latter series there is no hydrogen bonding within the system; furthermore, the overall morphology is more controllable than and considerably different from that of conventional urethane elastomers.

Another type of segmented copolymer that has been studied extensively by the dichroism techniques is that of dehydrohalogented poly(vinyl chloride) (DHPVC). These studies were initially undertaken by Shindo, Read, and Stein [53] and later extended by Wilkes et al. [54] to high-speed dichroism measurements. The polymer can be considered as segmented since local sequences of PVC monomer units have been dehydrohalogenated, thereby introducing a stiff polyene segment consisting of conjugated double bonds. The average length of

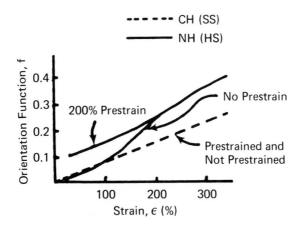

Figure 25. Plot of the orientation function vs. strain for the hard segment (HS) and soft segment (SS) of polyether-urethane elastomers that all had been initially strained to 200%. (Based on the data of Estes et al. [47].)

this stiff segment can be controlled by the time allowed for dehydrohalogenation.

Theoretical calculations of orientation behavior have been made based on a changing statistical segment length and associated absorbency. These calculations lead to the expression of the orientation function for a segment as

$$f = \frac{1}{15} \frac{(\overline{a_1 - a_2})}{\overline{A}_L} L \, (\lambda^2 - 1/\lambda), \tag{21}$$

where λ is the extension ratio,

$$(\overline{a_1 - a_2})_L = \frac{\displaystyle\sum_i (a_1 - a_2)_i N_i L_i^2}{\displaystyle\sum_i N_i L_i^2}, \tag{22}$$

and $\overline{A}_L = \Sigma_i N_i (a_1 + 2a)_i/3$. In Equations (21) and (22), N_i and L_i are the number and length of segments of type i in the network chain,* and the quantity $(a_1 - a_2)_i$ is the absorbence difference between the parallel and perpendicular directions to the chain axis for the ith segment. Comparison with experimental results is shown in Figure 26 for two very different average polyene lengths. The agreement is quite good up to fairly large deformation, while deviation oc-

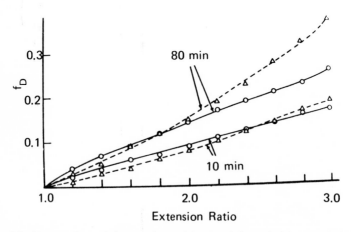

Figure 26. A comparison of the experimental (−−) and theoretical (−−) variations of the orientation function, f_D, with elongation for the 10-min- and 80-min-reaction-time DHPVC samples. (Courtesy of R. S. Stein [53].)

*This calculation is based on Gaussian rubber elasticity considerations; more general considerations are given by Flory and Abe [49] as well as Nagai [50].

curs to a greater extent for the longer statistical segment system. This greater deviation can be attributed to: (1) the expected deviation from Gaussian behavior as the number of statistical segments between network junctions becomes small; and (2) the fact that at the higher deformations, considerable orientation has occurred for the higher polyene material, which is not the case under typical Gaussian conditions.

To illustrate further the higher orientation with increasing statistical segment length (longer average polyene segment), Figure 27 shows the orientation function plots for a series of varying polyene lengths. Clearly, the orientation increases as length increases, as is predicted by theory. It is interesting to note that similar observations (also in agreement with theory) can be made for the birefringence on the same samples (see Figure 28). In the dichroism studies of Wilkes et al. [54] on these same polymers, it was shown that, for very rapid deformation rates, the dependence of orientation upon elongation was nonlinear, and there was considerable relaxation (disorientation) following this deformation. Furthermore, by using the polyene chromophore as an "orientation indicator," the degree and dependence of molecular orientation and relaxation upon plasticizer content of diethyl phthalate (DEP) was easily measured. This is illustrated in Figure 29 and, as expected, as the percentage of DEP increases, the degree of orientation decreases.

Little emphasis has been placed on the S–B–S block copolymers in dichroism

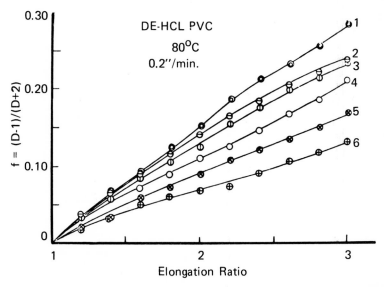

Figure 27. Plot of the orientation function for the polyene segments vs. elongation for samples prepared by different reaction times: (1) 160 min; (2) 80 min; (3) 40 min; (4) 20 min; (5) 10 min; (6) 5 min. Data measured at 80°C. (Courtesy of R. S. Stein [53].)

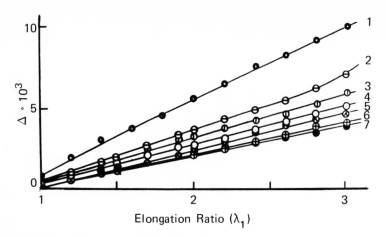

Figure 28. The variation in birefringence measured at 80°C with elongation for the DHPVC samples of polyenes having different segment lengths. Reaction times were (1) 160 min; (2) 80 min; (3) 40 min; (4) 20 min; (5) 10 min; (6) 5 min; (7) 0 min. (Courtesy of R. S. Stein [53].)

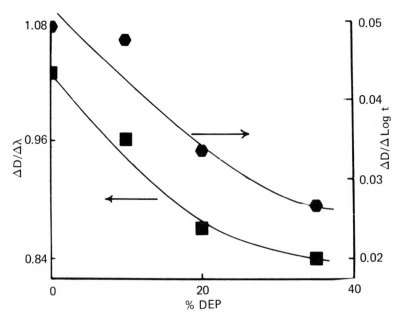

Figure 29. Effect of plasticizer (DEP) on the initial rate of change of polyene dichroism with elongation ratio and with time following elongation to 100%. (Courtesy of Wiley & Sons from [54].)

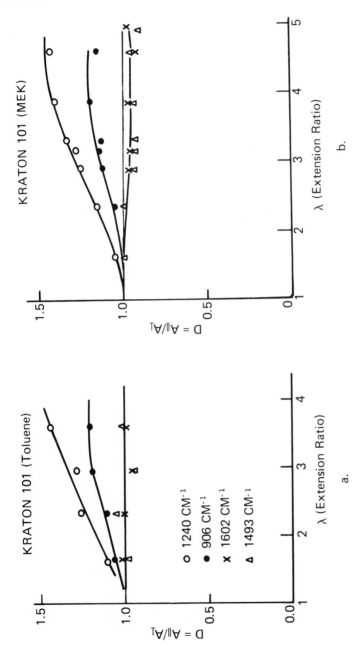

Figure 30. Plots of dichroism, D, versus extension ratio (λ) for four different absorption bands in an S–B–S (Kraton 101) block copolymer. The bands of 1,240 cm^{-1} and 906 cm^{-1} are associated with polybutadiene, while those of 1,493 cm^{-1} and 1,602 cm^{-1} are related to polystyrene: (a) film cast from toluene; (b) film cast from MEK.

measurements partially because of the experimental difficulty of separating absorption bands (in this as in other multicomponent systems) and therefore of being able to establish a baseline such that the absolute absorption can be determined leading to values of D and f. The degree of this difficulty, however, will depend on the frequency and extinction coefficients of the individual absorption bands of the components present. One example of the application of the dichroism method to the Kraton S–B–S was carried out by Read, Wilkes, and Stein [54]. This study was made as a function of extension ratios for the two types of solvent-cast film discussed earlier, i.e., films cast from toluene (rubbery) and from MEK (leatherlike). Figure 30 shows a comparison of the dichroism for four absorption bands. These data suggest that similar orientation behavior occurs over the time scale of the experiment and that there is little, if any, orientation of the PS phase (from a previous discussion, this lack of PS orientation is expected). It would be of interest to pursue similar studies as a function of time and temperature and of styrene content.

One last example is taken from the work of Onogi et al. on ethylene–propylene block copolymers [55]. Here dichroism was used to measure the orientation behavior of the c axis (chain axis) of the crystalline polypropylene component as a function of strain and propylene content. Their data, given in

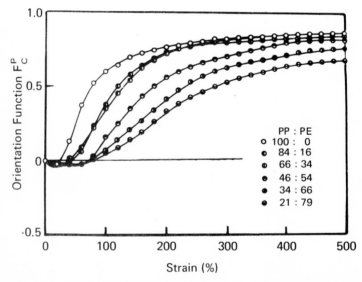

Figure 31. Plot of the orientation function for the c axis of the polypropylene crystal as a function of strain and of propylene content in a series of ethylene–propylene block copolymers. (Courtesy of S. Onogi [56].)

Figure 31, illustrate that the degree of orientation increases as the propylene content increases. The initial perpendicular orientation at strains less than 100 percent is typical of homopolymer behavior and therefore illustrates that within these block systems, some similarity in deformation mechanisms exists.

Photographic Small-Angle Light Scattering

The photographic SALS technique has been only recently extensively utilized in the copolymer area; specifically, it has been used to study segmented urethane systems that form anisotropic spherulites [4,57]. This technique, as explained earlier, is particularly suitable for aiding morphological characterization when optical anisotropic bodies are present in the sample, such as spherulites, rods, discs, etc. Figure 32 shows the type of urethane polymers studied by Samuels and Wilkes; in this series of four, the variable in chain structure is the number of sequentially located hard-segment units and, hence, hard-segment content. Figure 33 shows the H_v and V_v patterns as taken on a series of these polymers cast from chloroform at room temperature. The presence of a distinct spherulitic texture is indicated, and the average spherulite size, as calculated from the H_v patterns [4], depends on the hard-segment content as illustrated in Figure 34. Figure 35 shows the SEM micrographs of the surfaces of these films, and it is seen that the texture is spherulitic (note N4 in particular). Furthermore, the average spherulite size is in agreement with that calculated from the SALS pattern. By preparing films at different temperature conditions, these authors have shown that the spherulitic size can be increased and will display a more pronounced three-dimensionality, particularly in the members N2 and N4, as noted

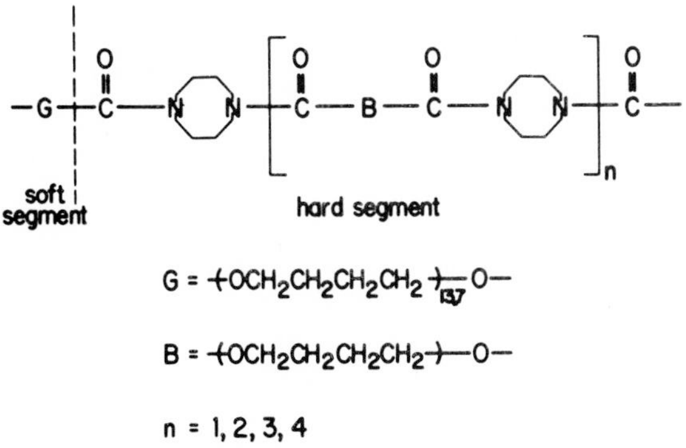

Figure 32. The series of segmented urethane polymers used by Samuels and Wilkes. Note that the number of hard-segment units is the variable.

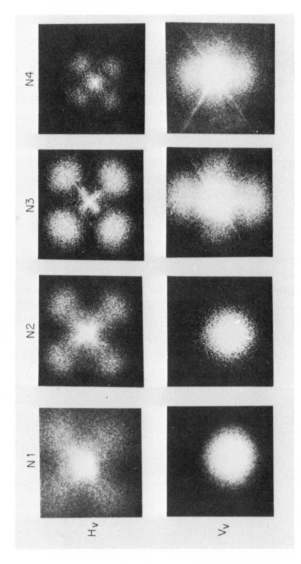

Figure 33. The H_v and V_v SALS patterns taken on the series of polymers shown in Figure 32. All films were cast from chloroform. The sample-to-film distance varies for the patterns. (Courtesy of Wiley & Sons from [4].)

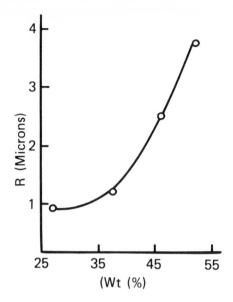

Figure 34. Plot of the size of the spherulite radius vs. hard-segment content for the series of polymers shown in Figure 32. (From [4].)

by SEM studies (see Figure 36). This porous yet elastomeric structure offers interesting potential applications as well as being an exceptional example of three-dimensional anisotropic spheres. The authors have indicated that this porous structure is formed as a result of the fact that the spherulites grow from solution and then precipitate out as the solvent is lost, which is in contrast to melt crystallization. Because of the "model" nature of the above systems, their deformation behavior has also been simultaneously studied by SALS and SEM. One example of this is shown in Figure 37, which gives the H_v scattering pattern and the corresponding SEM micrograph. The correlation between these SALS patterns, as based on the theory of spherulite deformations, and that of the actual deformation, is fair. At high deformation, however, there is no longer a uniform deformation (see Figure 38). This tends to lead to large void formation and an array of anisotropic geometries for the spherulites. Both of these affect the scattering and tend to eliminite a discrete pattern.

Based on the thermal dependence of the H_v scattering patterns, the patterns following deformation [57], and in correlation with optical and SEM studies [58], a model for the spherulitic structure has been proposed and is based on the fringe micelle concept rather than that of the usual folded-chain lamellar structure (see Figure 39).

The study of the above segmented urethanes is one example of how the initial

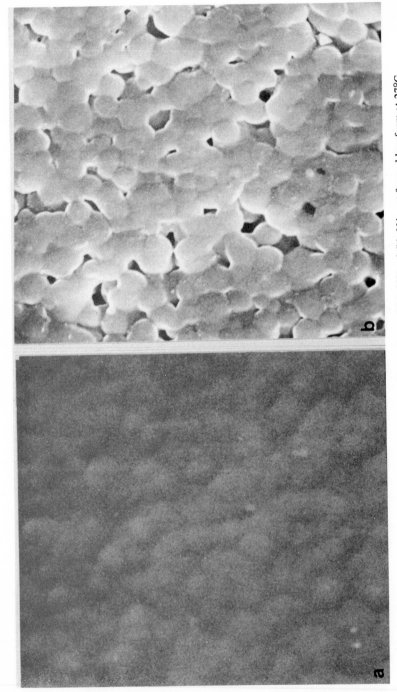

Figure 35. Scanning electron micrographs of the film surfaces of (a) N2 and (b) N4 cast from chloroform at 27°C.

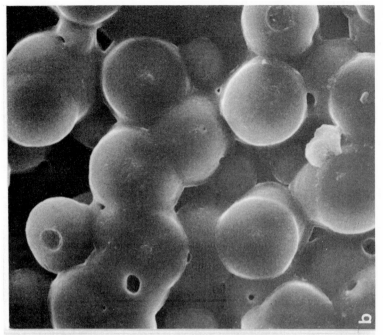

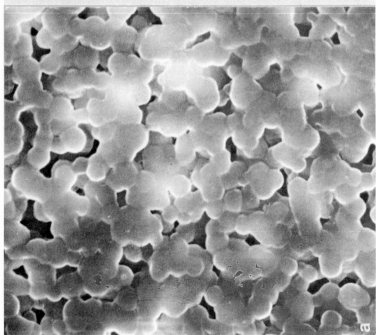

Figure 36. Scanning electron micrographs of the film surfaces of the segmented urethanes (a) N2 and (b) N4 cast from chloroform at 58°C.

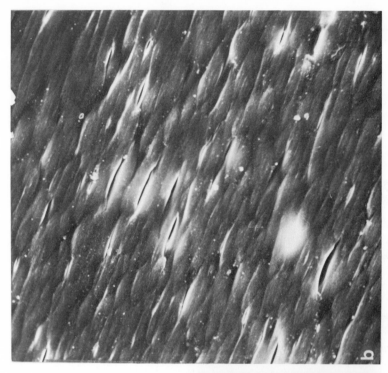

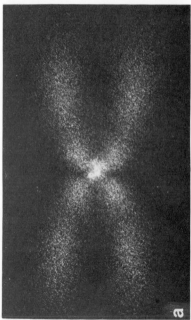

Figure 37. The H_v SALS pattern and corresponding scanning electron micrograph taken on a film of N3; elongation is $\sim 150\%$.

Figure 38. Scanning electron micrograph of the segmented urethane N4 as cast from chloroform at 58°C; elongation is about 200%.

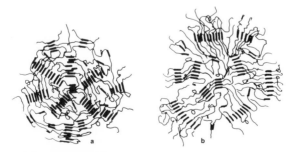

Figure 39. Two proposed models to account for the spherulitic structure present in the segmented urethane series N1–N4: (a) chain orientation tangential as hard segments crystallize radially; (b) chain orientation radial as hard segments crystallize tangentially.

application of the photographic method led to an extensive and most interesting investigation of segmented copolymers. It also illustrates the usefulness of correlating a rheo-optical method with other more common characterization methods like microscopy. Similar photographic SALS applications to the A–B or A–B–A semicrystalline block systems are limited at this time. From morphology studies, example systems would be the ethylene oxide–styrene A–B–A block polymers [59] as well as those based on the ethylene–propylene [56]. The latter are presently being investigated by the scattering technique [60].

Application of the photographic technique to the amorphous A–B or A–B–A systems also has been limited. The first studies were initiated by Stein and Wilkes [60] and concerned the S-B-S triblock systems. From cast films, H_v patterns (anisotropic scattering) were noted and displayed an azimuthal dependence. The pattern was not indicative of spherulitic-like scattering but showed a monotonic decrease in intensity with the radial angle. Upon sample deformation, this pattern deformed in a way similar to that arising from spherulites. Upon release of strain, the patterns were reversible, as shown in Figure 40. Azimuthal dependence of the V_v scattering occurred only upon deformation. The initial explana-

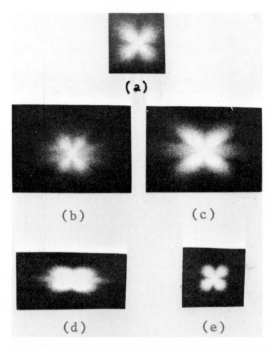

Figure 40. The H_v SALS patterns taken on an S-B-S block copolymer film cast from toluene at 40°C. Patterns were recorded as a function of percent elongation: (a) undeformed; (b) 40%; (c) 100%; (d) 175%; (e) after break.

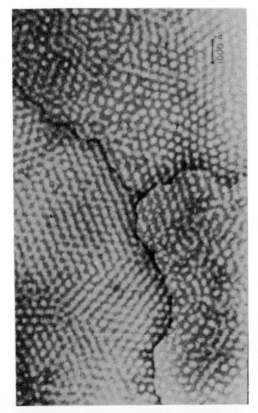

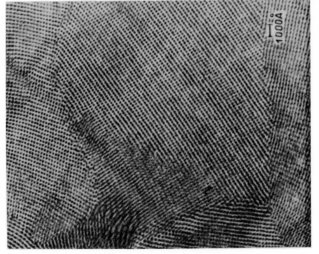

Figure 41. Transmission electron micrographs taken on two S–B–S block copolymer films that display the distinct grain boundaries suggested as the origin of H_v scattering. (Courtesy of M. Hoffmann [64].)

tion offered for these interesting results was based on scattering from the strain origin caused either by the way the styrene and butadiene precipitated out during film formation or by the localized strains induced in the rubbery phase as the last solvent was lost from the film. Later Stein [61] suggested that the anisotropic scattering might rather arise from form anisotropy effects or anisotropy associated with grain-boundary regions, which had been noted by Folkes and Keller [21], Lewis and Price [62], and Hoffman et al. [63,64] (See Figure 41). To further confirm this suggestion, Wilkes [65] observed the thermal dependence of the H_v patterns and noted that they remained or possibly increased in intensity upon heating well above the glass transition of system. Since it was expected that, had the scattering been of strain origin, it would be relaxed out when system flow could occur, the fact that the scattering was present suggested agreement with Stein's hypothesis. It should be mentioned that all of the scattering results were associated with films cast from toluene, which gives good phase separation, whereas no H_v patterns were observed from MEK-cast materials, which, as discussed earlier, results in a lower degree of phase separation.

Wide-Angle X-Ray Diffraction

Although small-angle X-ray diffraction has been used extensively in studying the structure of the block and segmented copolymers, there have been relatively few applications of the wide-angle technique. The prime reason for this is that the emphasis has been on the amorphous elastomers such as the S–B–S systems. However, several semicrystalline block or segmented systems exist, such as the EP blocks, the SEO blocks, and the segmented urethanes studied by Samuels and Wilkes [4], or those investigated by Bonart [50].

Earlier we presented one example of the application of wide-angle diffraction to the amorphous S–B–S system (see Figure 16). To illustrate its application to a semicrystalline system, we will use some of our own preliminary data on the segmented urethanes shown in Figure 32. Using the N3 (T_m = 145°C) polymer, we have uniaxially drawn a film to ~ 400 percent, recorded the diffraction pattern, and allowed the sample to relax (in hot water, 100°C), which resulted in a final permanent set of about 20 percent. The diffraction pattern was again recorded, and then the film was annealed without restraint at 20 degrees below the melting point to decrease further the degree of permanent set. Figure 42 shows the diffraction patterns from this experiment. It is noted that the initial undeformed pattern is well defined and, upon stretching to 400 percent, shows considerable orientation that can be related to the hard segment. After release of strain and heat-treatment in water, there is considerable orientation remaining, although the final elongation is only 20 percent (very little orientation is noted for an *initial* deformation to the same extension). In brief, these data provide further

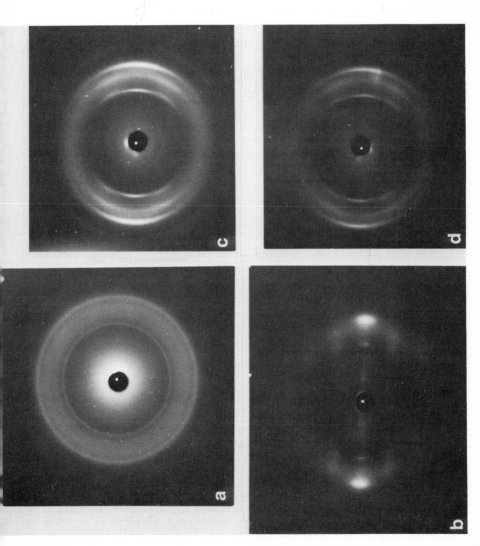

Figure 42. Wide-angle diffraction patterns of the N3 segmented urethane: (a) undeformed; (b) stretched to 400%; (c) released and allowed to relax in hot water; (d) after annealing without restraint at 20°C below the melting point.

evidence for the lack of reversibility in hard-segment orientation, as has been noted by Bonart [50] and by Estes et al. [46]. A difference exists between our segmented systems and those of other workers in that ours do not display hydrogen bonding. Therefore, formation of new hydrogen bonds cannot account for all of this permanent set orientation behavior, as has been suggested [50]. The details of our own work will be given elsewhere [57].

The above is only a single example of the usefulness of the wide-angle method. Clearly its application to other semicrystalline systems would be useful in studying component orientation.

Summary

Through numerous examples taken from the literature and from our laboratory, we have presented an introduction to the application of rheo-optical methods to the various types of block and segmented copolymeric materials. It is clear that rheo-optical methods have aided in developing an understanding of the structure–property relationships of these materials. Still, there are many unanswered questions. It is anticipated that these rheo-optical methods, used in conjunction with other characterization methods, will continue to help answer these questions and will aid in developing the ability to "custom build" polymeric systems to meet specific needs. A final point we have emphasized is the utility of applying more than a single method to gain information regarding structure and behavior. We believe that the multitechnique approach, although more demanding, leads to a better and more complete understanding of the relationship between structure and properties.

Acknowledgment

The authors wish to thank those writers and publishers cited in the captions for granting permission to reprint various figures from the literature.

References

1. Merrett, F.M., "Graft Polymers with Preset Molecular Configurations," *J. Polym. Sci.,* 24 (1957), 467.
2. Bradford, E.B. and Vanzo, E., "Ordered Structures of Styrene–Butadiene Block Copolymers in the Solid State," *J. Polym. Sci., Part A-1,* 6, no. 6 (1968), 1661.
3. Hendus, Von H., Illers, K.H., and Ropte, E., "Struk turuntersuchungen on Styrol-Butadien-Styrol Blockcopolymeren," *Kolloid Z.-Z. Polym.,* 216/217 (1967), 110.
4. Samuels, S. and Wilkes, G., "Anisotropic Superstructure in Segmented Polyurethanes as Measured by Photographic Light Scattering," *J. Polym. Sci., Part B,* 9 (1971), 761.

5. Kuhn, W. and Grün, F., "Beziehung zwischen elastichen Konstanten und Dehnungs-Dappelbrechung hochelasticker Stoffe," *Kolloid Z.-Z. Polym.*, 101 (1942), 248.

6. Wilkes, G., "The Measurement of Molecular Orientation in Polymeric Solids," *Adv. Polym. Sci.*, 8 (1971), 91.

7. Stein, R.S., Onogi, S., and Keedy, D.A., "The Dynamic Birefringence of High Polymers," *J. Polym. Sci.*, 57 (1962), 801.

8. Erhardt, P.F. and Stein, R.S., "Rheo-optics of High Speed Deformation [of Polymers]," *Appl. Polym. Symp.*, no. 5 (1967), 113.

9. Fraser, R.D.B., "The Interpretation of Infrared Dichroism in Fibrous Protein Structures," *J. Chem. Phys.*, 21, no. 9 (1953), 1511.

10. Alexander, L.E., *X-Ray Diffraction Methods in Polymer Science*, New York: Wiley (1969).

11. Guinier, A., *X-Ray Diffraction in Crystals, Imperfect Crystals and Amorphous Bodies*, San Francisco: W.H. Freeman (1963).

12. Bunn, C.W., *Chemical Crystallography*, Oxford: Clarendon Press (1946).

13. Samuels, R.J., "Morphology of Deformed Polypropylene. Quantitative Relations by Combined X-ray, Optical, and Sonic Methods," *J. Polym. Sci., Part A*, 3 (1965), 1741.

14. Stein, R.S. and Oda, T., "Direct Observation of the Orientation Rate of Polyethylene Crystals," *J. Polym. Sci., Part B*, 9 (1971), 543.

15. Kawaguchi, T., "Principles of Dynamic X-ray Diffraction Technique," University of Massachusetts, Polymer Research Institute, Amherst, Mass., Office of Naval Research Contract Report, ONR-TR-86, April 1966 (AD 644-029).

16. Stein, R.S., Erhardt, P., Clough, S., Van Aarsten, J.J., and Rhodes, H.B., "The Scattering of Light by Solid Polymer Films Having Partially Ordered Structure," 2d Interdisciplinary Conference on Electromagnetic Scattering, University of Massachusetts, 1965, *Electromagnetic Scattering*, R.L. Rowell and R.S. Stein, eds., New York: Gordon and Breach (1967), 339.

17. Desper, C.R. and Kimura, I., "Mathematics of the Polarized-Fluoresence Experiment," *J. Appl. Phys.*, 38 (1967), 4225.

18. Nishijima, Y., Ogoni, Y., and Asai, T., "Fluoresence Method for Studying Molecular Orientation in Polymer Solids," *Rep. Prog. Polym. Phys. Jap.*, 8 (1965), 131.

19. Snyder, R.G., "Raman Scattering Activities for Partially Oriented Molecules," *J. Mol. Spectrosc.*, 37 (1971), 353.

20. Beecher, J.F., Marker, L., Bradford, R.D., and Aggarwal, S.L., "Morphology and Mechanical Behavior of Block Polymers," *J. Polym. Sci., Part C*, 26 (1969), 117.

21. Folkes, M.J. and Keller, A., "The Birefringence and Mechanical Properties of a 'Single Crystal' from a Three Block Copolymer," *Polymer*, 12 (1971), 222.

22. Dlugosz, J., Keller, A., and Pedemonte, E., "Electron Microscope Evidence of a Macroscopic 'Single Crystal' From a Three Block Copolymer," *Kolloid Z.-Z. Polym.*, 242 (1970), 1125.

23. LeGrand, D.C., "Domain Formation in Block Copolymers," *J. Polym. Sci., Part B*, 8 (1970), 195.

24. Clough, S.B., Schneider, N.S., and King, A.D., "Small-Angle X-ray Scattering From Polyurethane Elastomers," *J. Macromol. Sci., Phys.*, B2, no. 4 (1968), 641.

25. Kaelble, D.H., "Interfacial Morphology and Mechanical Properties of ABA Triblock Copolymers," *Trans. Soc. Rheol.*, 15, no. 2 (1971), 235.

26. Samuels, S. and Wilkes, G., unpublished results.

27. Leary, D.F. and Williams, M.C., "Statistical Thermodynamics of ABA Block Copolymers: I," *J. Polym. Sci., Part B*, 8, no. 5 (1970), 335.

28. Wilkes, G. and Stein, R.S., unpublished results.

29. Inoue, T., Moritani, M., Hashimoto, T., and Kawai, H., "Deformation Mechanism of Elastomeric Block Copolymers Having Spherical Domains of Hard Segments Under Uniaxial Tensile Stress," *Macromolecules,* 4, no. 4 (1971), 500.
30. Moritani, M., Inoue, T., Motegi, M., and Kawai, H., "Light Scattering From a Two-Phase Polymer System. Scattering From a Spherical Domain Structure and Its Explanation in Terms of Heterogeneity Parameters," *Macromolecules,* 3, no. 4 (1970), 433.
31. Debye, P. and Bueche, A.M., "Scattering by an Inhomogeneous Solid," *J. Appl. Phys.,* 20 (1949), 518.
32. Goldstein, M. and Michalek, E.R., "Theory of Scattering by an Inhomogeneous Solid Possessing Fluctuations in Density and Anisotropy," *J. Appl. Phys.,* 26 (1955), 1450.
33. Stein, R.S. and Wilson, P.R., "Scattering of Light by Polymer Films Possessing Correlated Orientation Functions," *J. Appl. Phys.,* 33 (1962), 1914.
34. Wilkes, G.L. and Marchessault, R.H., "Light Scattering Characterization of Polyvinyl Acetate Latex Films," *J. Appl. Phys.,* 37 (1966), 3974.
35. Keijzers, A.E.M., Van Aartsen, J.J., and Prins, W., "Light Scattering by Crystalline Polystyrene and Polypropylene," *J. Amer. Chem. Soc.,* 90 (1968), 3107.
36. Stein, R.S., Erhardt, P.F., Clough, S.B., and Adams, G., "Scattering of Light by Films Having Non-random Orientation Fluctuations," *J. Appl. Phys.,* 37 (1966), 3980.
37. Clough, S.B. and Schneider, N.S., "Structural Studies on Urethane Elastomers," *J. Macromol. Sci., Phys.,* B2, no. 4 (1968), 553.
38. Wilkes, G.L. and Stein, R.S., "Effect of Morphology on the Mechanical and Rheo-optical Properties of a Styrene–Butadiene–Styrene Block Copolymer," *J. Polym. Sci., Part A-2,* 7 (1969), 1525.
39. Fischer, E., unpublished data.
40. Nielson, L.E. and Buchdahl, R., "Viscoelastic and Photoelastic Properties of Polystyrene Above Its Softening Temperature," *J. Chem. Phys.,* 17 (1949), 839.
41. Andrews, R.D. and Rudd, J.F., "Photoelastic Properties of Polystyrene in Glassy State. I. Effect of Molecular Orientation," *J. Appl. Phys.,* 28 (1957), 1091.
42. Fischer, E. and Henderson, J.F., "The Stress–Strain–Birefringence Properties of Styrene–Butadiene Block Copolymers," *J. Polym. Sci., Part C,* 26 (1969), 149.
43. Takayanagi, M., Uermura, S., and Minami, S., "Application of Equivalent Model Method to Dynamic Rheo-Optical Properties of Crystalline Powder," *J. Polym. Sci., Part C,* 5 (1963), 113.
44. Wiener, O., "Die Theorie des Mischkerpers fur das Feld der stationären Stromung," *Sachs. Ges. Wiss. Leipzig. Math.-Phys.,* Kl Akh 32 (1912), 509.
45. Abronn, H. and Frey, A., *Das Polarisationmikroskop,* Leipzig: Academische Verlagsgellschaft M.H.B (1926).
46. Estes, G.M., Seymour, R.W., and Cooper, S.L., "Infrared Studies of Segmented Polyurethane Elastomers. II. Infrared Dichroism," *Macromolecules,* 4 (1971), 452.
47. Shindo, Y. and Stein, R.S., "The Birefringence and Dichroism of Rubbery Copolymers," *J. Polym. Sci., Part A-2,* 7 (1969), 2115.
48. Flory, P.J. and Abe, Y., "Preferential Orientation and Strain-Dichroism of Polymer Chains," *Macromolecules,* 2 (1969), 335.
49. Nagai, K., "Stress-Optical and -Dichroic Coefficients of Polymeric Networks," in *Progress in Polymer Science,* Vol. 1, H. Imoto and S. Onogi, eds., Tokyo: Kodansha (1971).
50. Bonart, R., "X-ray Investigations Concerning Physical Structure of Cross-linking in Segmented Urethane Elastomers," *J. Macromol. Sci., Phys.,* B2, no. 1 (1968), 115.
51. Cooper, S.L., private communication.

52. Shindo, Y., Read, B.E., and Stein, R.S., "The Study of the Orientation of Polyvinyl Chloride Films by Means of Birefringence and Infrared, Visible and Ultraviolet Dichroism," *Makromol. Chem.,* 118 (1968), 272.

53. Wilkes, G.L., Uemura, Y., and Stein, R.S., "Apparatus for Instantaneously Measuring Ultraviolet, Visible, or Infrared Dichroism from Thin Polymer During High-Speed Stretching," *J. Polym. Sci., Part A-2,* 9 (1971), 2151.

54. Read, B., Wilkes, G., and Stein, R.S., "Polarized Infrared Studies of a Styrene–Butadiene Block Copolymer," unpublished status report, University of Massachusetts, Polymer Research Institute, Amherst, Mass., 1968.

55. Onogi, S., Asada, T., and Sakai, K., "Rheo-optical Studies of High Polymers. XVI. Infrared Dichroic and Viscoelastic Behavior of Ethylene–Propylene Block Copolymers," *Proceedings of the Fifth International Conference on Rheology,* Vol. 4, Kyoto, October 7–11, 1968, S. Onogi, ed., Tokyo: University of Tokyo Press, and Baltimore, Md.: University Park Press (1970), 41.

56. Samuels, S. and Wilkes, G., "Further Studies on the Superstructure in a Series of Hand-Soft Segmented Urethanes," *Polym. Prepr., Amer. Chem. Soc., Div. Polym. Chem.,* 13, no. 2 (1972), 999.

57. Samuels, S. and Wilkes, G., in preparation.

58. Short, J. and Crystal, R., "Morphology of Block Copolymers," *Appl. Polym. Symp.,* no. 16 (1971), 137.

59. Stein, R.S., private communication.

60. Stein, R.S. and Wilkes, G.L., "Scattering of Light from Deformed Regions Surrounding Voids and Inclusions in High Polymers," *J. Polym. Sci., Part A-2,* 7 (1969), 1695.

61. Stein, R.S., "On Depolarized Light Scattering from Block Copolymers," *J. Polym. Sci., Part B,* 9 (1971), 747.

62. Lewis, P.R. and Price, C., "Morphology of ABA Block Polymers," *Nature (London),* 223 (1969), 494.

63. Kampf, G., Krömer, H., and Hoffman, M., "Long Range Order of Supramolecular Structures in Amorphous Butadiene–Styrene Block Copolymers," *J. Macromol. Sci., Phys.,* B6, no. 1 (1972), 167.

64. Kampf, G., Krömer, H., and Hoffman, M., "Weitreichende ordnungen der ubermolekularen Strukturen in amorphen Butadiene-Styrol-Blockcopolymeren," *Kolloid Z.-Z. Polym.,* 247 (1971), 820.

65. Wilkes, G.L., "Comments on Light Scattering from Anisotropic Regions in ABA Block Copolymers," *J. Polym. Sci., Part A-2,* 10 (1972), 767.

SPECIAL APPLICATIONS

MODERATOR: VIVIAN STANNETT
Department of Chemical Engineering
North Carolina State University
Raleigh, North Carolina

13. Special Properties of Block and Graft Copolymers and Applications in Film Form

VIVIAN STANNETT

North Carolina State University
Raleigh, North Carolina

ABSTRACT

Some special properties that distinguish block and graft copolymers from normal random copolymers are discussed, including compatibility, domain structures, and conformational aspects. Actual and potential applications of block and graft copolymers in film form are outlined. These include high-barrier films and coatings and membranes for separation procedures including reverse osmosis. Other applications of surface grafting for biomedical and other uses due to improved abrasion resistance, compatibility, and adhesion to films are briefly described.

Block and graft copolymers have a number of special properties that make them quite different from homopolymers and random copolymers [1]. In some ways block and graft copolymers can be treated together, particularly since difficulties of synthesis often limit block copolymers to only one or two blocks of each component, and graft copolymers to only one or two grafted side chains per backbone. When block copolymers consist of many short blocks, however, their properties begin to approach those of random copolymers rather than graft copolymers.

Graft copolymers and single or double block copolymers are unique in that in many ways the properties of the individual polymers are maintained together with those of the other polymeric component. In other words, block and graft copolymerizations give an opportunity to impose the properties of one polymer on a second polymer without greatly changing those of the main constituent. Thus, up to about 15 percent acrylic acid can be grafted to high-density polyethylene or nylon without reducing the modulus, softening point, or mechanical properties. At the same time the nylon and polyethylene have high water sorption and antistatic properties due to the grafted side chains. In the case of polyethylene, the grafted polymer has superior adhesion properties and will tolerate the addition of fillers and form films that can be heat-sealed to aluminum [2,3]. This particular feature of graft copolymers has led to extensive efforts to modify natural and synthetic fibers to import permanent soil-release, antistatic, dyeability, and other properties to textile fibers.

Other special properties of graft and block copolymers that lead to potential or actual applications are the changes in solubility and compatibility, the mosaic structure of such polymers in the solid state, and the possibility of conformational changes. These and other special properties are considered. Finally, the possible applications of graft and block copolymers in film form are discussed.

Some Special Properties of Practical Interest

Solubility and Conformations

The solubility of block and graft copolymers is of considerable practical interest. Block and graft copolymers are often soluble in mixtures of solvents for each component but not in either solvent alone. There is a threshold solubility composition at each end of the scale; at these points the intrinsic viscosities are at a minimum and rise to a maximum at some intermediate solvent composition. Films cast at close to the two minimum-viscosity compositions often display properties closer to the corresponding homopolymers, thus allowing some anisotropy to be developed in the films. A block or graft consisting of a glassy plus a rubbery polymer component can therefore be made to behave as a glass or a rubber by casting in different solvents or by treating the films with the appropriate solvents after formation. Similarly, attempts have been made to introduce anisotropy across membranes by this technique.

Another practical consequence of block or graft formation is that it introduces considerable tolerance in the polymer towards nonsolvents. Thus, a styrene/secondary-cellulose-acetate graft copolymer in a solvent will tolerate large additions of toluene or similar hydrocarbons. Such characteristics can be well utilized, for example, in formulating adhesive or coating compositions.

Compatibility

An important practical feature of simple block and graft copolymers is their ability to make compatible two otherwise incompatible polymers. Cellulose acetate is highly incompatible with polystyrene, and even a few percent of one polymer mixed with the other gives white opaque films on casting from solvents. The addition of a graft copolymer of the two, however, in comparatively low percentages, causes compatibilization in the sense that clear transparent films are formed [4]. In Figure 1 is shown a plot of light transmission of films of a 50:50 mixture by weight of secondary cellulose acetate and polystyrene in which increasing amounts of a graft copolymer, consisting of roughly equal proportions of cellulose acetate and polystyrene, have been blended. It can be seen that the excellent clarity of the cast pure graft film is maintained up to about 60 percent of the 50:50 mixture, and good clarity up to about 80 percent. The complete three-component phase diagram, with 75 percent light transmission as an arbitrary measure of compatibility, is shown in Figure 2. Again, it can be seen that the graft copolymer has succeeded in making compatible a mixture of polystyrene and cellulose acetate over a very wide range of three-component compositions. In the same study, however, it was also shown that grafts with short side chains or backbones did not give a large degree of compatibilization, although blends of each type did.

Microdomain Structure

In the above discussion, the criterion for compatibility was the clarity of films cast from the various polymer blends. There is evidence, however, that even films cast from pure graft copolymers consist, in reality, of microdomains

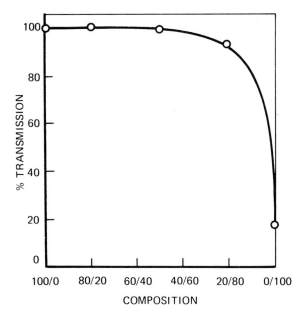

Figure 1. Light transmission vs. composition of films made from a 50:50 blend of secondary cellulose acetate and polystyrene and a 44.1% polystyrene-grafted cellulose acetate.

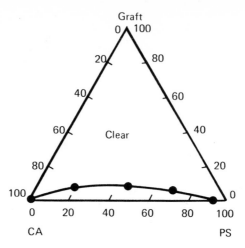

Figure 2. Phase diagram of blends of secondary cellulose acetate, polystyrene, and a 44.1% polystyrene-grafted cellulose acetate. A 75% light transmission of the film was selected as an arbitrary compatibility limit.

of each type of polymer. Thus, for example [4], the diffusion constant for water at low activities in cellulose acetate is 1.8×10^{-8} cm^2/sec, compared with 28.0×10^{-8} cm^2/sec for polystyrene. The diffusion for water in graft copolymer containing 44.1 percent combined polystyrene, however, was found to be only 3.4×10^{-8} cm^2/sec. If the graft copolymer film were molecularly homogeneous, one would expect the diffusivity of the graft to be approximately intermediate between those of the two homopolymers, i.e., about 13×10^{-8} cm. On the other hand, if the graft copolymer film consisted of isolated domains of each polymeric species, the diffusivity would be governed by the slow diffusion step through the cellulose acetate domains. Since the volume fraction is about 50 percent, the value would be 3.6×10^{-8} cm^2/sec, in good agreement with the measured value of 3.4.

The implications are clearly that the graft copolymer films are not molecularly homogeneous but consist essentially of domains of each type of polymer. The fact that the films are quite clear is an indication that the domains are extremely small. Considerable evidence for a similar structure in block copolymers has been advanced [5,6]. The compatibility aspects of graft copolymers may have to be somewhat modified in the face of such findings. For example, it may be that graft or block copolymers on their own form clear, apparently homogeneous films. However, the evidence suggests that these consist of very small domains of each polymeric species. Recent evidence in the case of block suggest only about 50-1000 Å size for the domains [6]. Presumably, when a given homopolymer is added to a graft or block copolymer, it can blend with the domain of its own kind up to the point when the average domain is large enough to scatter light; it then begins to show opacity. An extreme of this situation would be when only a small amount of graft copolymer is available, but sufficient to form a compatible interface. It now seems clear that the success of blends of opaque rubbery and glass polymers, such as high-impact polystyrene and A-B-S polymers, is partly due to the deliberate of fortuitous presence of compatibilizing graft copolymers in the blend.

Effects of Grafting on Gas Vapor and Liquid Transport

In principle, grafting is an attractive way of changing the permeability characteristics of plastic films. The appropriate monomer could be sorbed into the substrate film and rapidly polymerized under an electron beam, for example. Homopolymer probably need not be removed, because the graft copolymer formed would lead to compatibility. Many attractive alternative methods of grafting are also available.

Early work [7] on grafting styrene and acrylonitrile to polyethylene films showed a progressive decrease in gas permeability with grafting. Subsequently, more detailed work by Huang and Kanitz [8,9] with a similar system showed that as the extent of grafting styrene to polyethylene film increased, the permeability first decreased and then began to increase again. This was attributed to a decrease in the free volume with lower degrees of grafting and a disruption of the crystallinity at higher degrees of grafting. The densities of the films were also measured and did indeed show an increase slightly above the additive value in the early stages. Interestingly, acrylonitrile grafts did not show a minimum but only a progressive decrease in permeability. Since the grafted polyacrylonitrile chains are not even swollen in the grafting solution, they are probably inefficient in causing a disruption of the crystallinity. In the styrene system, on the other hand, progressive swelling of the film leads to disruption of the crystallinity.

A direct demonstration of these effects can be seen in the recent work of Williams and Stannett [10] with polyoxymethylene films. Butadiene and acrylonitrile were radiation-grafted onto films, each with 18 percent dimethyl formamide as a swelling agent in the monomer. The grafting-time curves are shown in Figure 3. It is clear that butadiene, which swells its own polymer, leads to a progressive increase in grafting with time. Acrylonitrile, however, grafts as a hole-filling type of reaction, and the grafting rapidly reaches a maximum and then levels off. The water vapor permeability of two grafted films is shown as a function of humidity in Figure 4. The results are quite startling; polyoxymethylene films containing less than 10 percent grafted polymers showed a twofold increase in the water vapor permeability with butadiene grafting and a sixfold *decrease* with acrylonitrile grafting.

Somewhat similar changes in the transport behavior of polymer films with

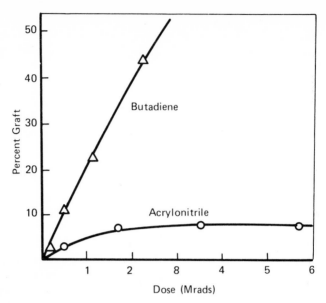

Figure 3. Percent graft vs. dose curves for butadiene and acrylonitrile to polyoxymethylene film. The dose rate was 0.1 Mrads/hr at 25°C. Both monomers contained 18% dimethyl formamide.

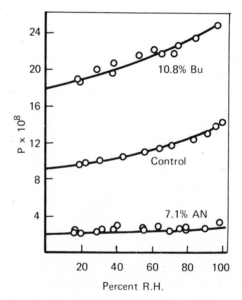

Figure 4. Water vapor permeability $(ccs/cm^2/cm/sec/cm \ Hg)$ vs. relative humidity for grafted and ungrafted polyoxymethylene films.

grafting were shown in recent work on grafting styrene to cellulose acetate membranes for use in reverse osmosis applications [11–13]. The mutual radiation grafting of styrene gave a curve of the same shape as that of the acrylonitrile example in Figure 3. It was found to be impossible to increase the grafting above the plateau level. Finally, on completely dissolving the film and recasting it was possible to regraft in a continually increasing fashion, as shown in Figure 3 with butadiene. A separate study of the swelling of the films in the 90 percent styrene–10 percent pyridine grafting solution indicated that the "plateau"-grafted films showed almost no swelling and correspondingly little grafting. The recast films, on the other hand, swelled more than the original cellulose acetate film and grafted at a correspondingly faster rate.

It is believed that the small but definite degree of crystallinity in the cellulose acetate films exerts a restraining influence on the grafting. The grafting reaction proceeds until the film has become "filled" with polystyrene. Under these conditions it cannot swell or graft. On dissolving and recasting, the grafted side chains prevent crystallization, and the film is free to continue grafting and expanding indefinitely. The water transport properties are given in Table I. It can be seen that the original grafted polymer had a reduced flux and a higher density than calculated, whereas the recast film was bulkier and had a higher flux.

In grafting directly [13] in a monomer solution containing carbon tetrachloride as a chain-transfer agent, the grafting rate curves lost their plateau character and grafted similar to the butadiene case (Figure 4). The water transport properties of such a film are included in Table I. Greatly increased water transport can be seen; the salt rejection, on the other hand, dropped only from 99.2 to 98.5 percent [13]. In all cases, the grafted films showed greatly reduced wet creep behavior, a property believed to be of importance to Loeb-type membrance technology.

These few examples illustrate how grafting technology can be used to modify the "barrier" properties of films, in the direction of both decreased and in-

TABLE I

Water Transport Properties of Styrene-Grafted Cellulose Acetate Films (35% graft)

Film	Water Regain* (%)	Diffusivity* ($cm^2/sec \times 10^8$)	Flux† (gfd-mil)	Density
C. A. film	4.5	2.5	0.035	1.285
Plateau graft	2.3	2.1	0.006	1.223‡
Recast from dioxane	3.5	1.8	0.017	1.202‡
Short-chain grafted	6.9	3.1	0.23	—

*25°C, 50% H.
†800 psi, 0.5% NaCl.
‡Calculated density = 1.217.

creased permeabilities. It is hoped that studies of this kind will lead to practical applications.

Block Copolymers

The phase separation of block copolymers can be utilized favorably in many aspects of polymer films. Since the properties such as permeabilities and viscoelastic characteristics of each domain (phase) of the block copolymer films are, for the most part, close to those of the homopolymer representing the block sequence, overall film properties depend on the combination of type of polymers incorporated in each block.

The major distinction and, perhaps, advantage of block copolymers over random and graft copolymers is the fact that the properties of each polymer phase are retained. Consequently, the advantageous properties of two polymers can be combined in a more favorable manner. For instance, poly(dimethyl siloxane) has very high permeabilities to gases and vapors, but because of its rubbery nature, it is difficult to fabricate into a thin film. Block copolymers of dimethyl-siloxane and polycarbonate combine the high permeabilities of polydimethyl-siloxane and the mechanical properties of the highly glassy polymer. The block copolymers can be easily cast or spun into hollow fibers, and the improvement of viscoelastic properties is more than enough to compensate for the marginal reduction in overall permeabilities from those for poly(dimethyl siloxane).

Another example of utilization of phase separation can be seen in examples of mosaic membranes. Mosaic membranes of hydrophilic domain and hydrophobic domain provide the advantages of highly swollen hydrophilic polymers with the ease in fabrication and handling of hydrophobic polymers. Examples of this type of mosiac membrane are block copolymers of polyethyleneoxide–poly-ethylene, sulfonated polybutadiene-polystyrene, poly (vinyl pyridinum salt)-polystyrene, etc.

Charge mosaic membranes, which consist of discretely separated positively-and negatively-charged domains, may draw special interest as piezodialysis (pressure dialysis) membranes. In charge mosaic membranes, each domain permits the passage of mobile counterions and enhances the transport of ionic solutes when a solution is forced to pass through the membrane by applied pressure. In single-charged ionic polymer film, the counter-mobile ion can go through the film, but because of the streaming potential created by the imper-meability of mobile co-ions, the net transport of whole salt (cations and anions) is greatly reduced. In this respect, the charge mosaic membrane provides the short circuit to nullify the streaming potential.

The practicability of charge mosaic membranes of block copolymers as piezo-dialysis membranes intended to enrich salt solutions is somewhat doubtful because of the following factors: (1) the size of domain obtainable with block copolymers is generally too small; (2) the introduction of two charges generally yields too great swelling unless the third nonionic domain is utilized as a cross-link (by using A-B-C-type block copolymers); and (3) poly–poly salt formed between two phases seems to negate the piezodialysis effect. Although they may not be satisfactory for desalting purposes, the high selectivities of such charge mosaic membranes for ionic and nonionic solutes may provide some unique applications in membrane processes.

Surface Grafting by Plasma Polymerization

It has been known for many years that some organic compounds form polymers in plasma, though the polymers were recognized as by-products of phenomena associated with electric discharge. It was only relatively recently (about the 1960s) that this phenomenon was utilized practically to make a special coating on metal.

Once some of the advantageous features of plasma coating (e.g., flawless thin coatings of polymers, unique properties such as low dielectric constant, and good corrosion resistance) were recognized, much applied research concerning the use of the process was done. Perhaps because of the interest in electrical properties of deposited polymer film, most of the studies appearing in the literature are concerned with polymer deposition onto metals that were used as electrodes in a low-pressure glow discharge. Although the reactions (polymer as a by-product) of organic compounds in an electrodeless (inductive coupling) glow discharge and the properties of polymers formed by an electrodeless (inductive coupling) glow discharge have also been treated in the literature, neither the detailed polymerization mechanism nor the structure of polymer as formed by electrodeless (inductive coupling) glow discharge had been investigated.

Glow discharge polymerization can be achieved by a number of methods, e.g., d.c. and a.c. discharge with electrodes, and electrodeless discharge with a.c. (radio frequency and microwave) discharge. The polymer deposition pattern (i.e., the physical location of polymer deposition, polymer deposition rates, and the kind of polymer deposit) is highly dependent upon the type of discharge and the geometrical aspect of a reactor (discharge vessel). Although it is difficult to generalize, the following trends can be drawn from results obtained by various types of discharge dealt with in the literature.

1. In d.c. discharge, the deposition of polymer takes place almost exclusively on the cathode, which indicates that positive ions formed in glow discharge play an important role in plasma polymerization.

2. In a.c. discharge with electrodes, polymer is deposited on the surface of

both electrodes, but almost exclusively on the electrode surface. When a flow system is employed, polymer is deposited on the wall of the reaction vessel on the downstream side of flow at a very high flow rate. Electrodeless glow discharge with capacitive coupling (electrodes on the outside of the reaction vessel) can be considered similar to electrode discharge. In this case, polymer is deposited on the wall of the reaction vessel, but the deposition seems to be limited to the region between the two electrodes and slightly on the downstream side.

3. In electrodeless discharge (with inductive coupling), the discharge tube is placed inside a solenoid through which an alternating current is passed. A discharge will be established when the current and frequency are sufficiently high. This type of discharge is referred to as "electrodeless discharge" in this paper. The electrodeless discharge requires higher frequency than electrode discharge, and either radio frequency or microwave is generally used. In electrodeless discharge, polymer is deposited on the wall of the reaction vessel in the glow region, with exceptionally heavy deposition on the portion in the solenoid. In a flow system, polymer is also deposited on the wall in the nonglow region on the downstream side. The advantage of electrodeless glow discharge is that polymer can be deposited on an exposed surface of any shape (without the restriction caused by the shape and size of an electrode as in the former case).

Insofar as the approach to the modification of a surface is concerned, the effects of the glow discharge are somewhat similar to those of surface grafting. Therefore, some important aspects of glow discharge polymerization are compared with conventional grafting so that advantages and limitations of both methods can be noted.

1. *Thickness of modified surface.* The most significant difference between a surface modified by conventional grafting and one modified by plasma polymerization is the thickness of the modified layer. Although no clear definitions are made, the term "surface grafting" is used in contrast to the homogeneous grafting in which grafting extends all the way into the bulk of the original polymer sample. The thickness resulting from surface grafting is on the order of 10,000 to 100,000 Å (1 to 10 μ).

Typical deposition thicknesses by plasma polymerization are on the order of 100 to 2,000 Å. Although it is possible to build up the thickness beyond this range, it has been found that the characteristic properties of the deposited polymer change with increased thickness. Stronger bonding and more uniform surface coverage can be obtained with deposition in the lower range. In this range the macroscopic texture of the surface is not altered, though all the surface is uniformly covered by the polymer deposition. This implies that it is possible to cover a surface with a new material without changing its bulk property and macroscopic surface texture. For instance, if silicone is deposited by glow discharge on a silicone rubber surface that has an exposed surface of filler pigments, the modified surface is completely covered by the silicone polymer.

2. *Dependence upon the nature of substrate polymer and the reactivity of monomers.* These two factors are vitally important to a comparison of the efficiency of conventional grafting and glow discharge polymerization. In conventional grafting, the efficiency of grafting copolymerization is highly dependent upon the nature of the polymers and monomers used. A particular monomer may graft well onto a particular polymer, but the same monomer may not graft at all onto another polymer. This is particularly true for chemical means of graft initiation. Even with the less specific radiation-initiated graft copolymerization (mutual irradiation of polymer and monomer), the efficiency depends on the free-radical yields (G values) of polymer and monomer. Only in those cases where the G value of the polymer is high and the G value of the monomer is low will effective grafting be obtained. Consequently, one monomer may graft well onto one polymer, but the same monomer may not graft at all onto another polymer, even with radiation grafting. This is also true of the preirradiation method and of the peroxide method used in radiation-initiated graft copolymerization.

In glow discharge polymerization, the deposition of polymer is not appreciably affected by the nature of the substrate polymer. A polymer deposits equally well on the surface of glass, polymer, or metal. Therefore, once a desirable monomer has been found, the method could be used to apply it to any type of polymer regardless of chemical nature, morphology, or presence of fillers.

The physical and chemical properties of the resultant surface would depend on the chemical nature of the monomer; however, special caution should be taken in predicting the surface properties from the chemical structure of the monomer used. Some functional groups of monomers have been found to cause decomposition of the monomer in a glow discharge and not remain in the deposited polymer. Some functional groups cause decomposition of the monomer but do remain in the deposited polymers.

Plasma-polymerized polymers can also be used as polymer films by deposition on a porous substrate. In this approach unique composite membranes of ultrathin film tightly bonded to substrate polymer (by grafting) can be prepared.

Electrodeless glow discharge polymerization has been extensively studied by Dr. H. Yasuda at Camille Dreyfus Laboratory, Research Triangle Institute, to prepare membranes for use in desalination by reverse osmosis. Excellent results were obtained in forming ultrathin semipermeable membranes on the surface of porous substrates such as porous polymer membranes and porous glass tubes. Some of the results are shown in Table II. Improved barrier properties can also be imparted to polymer films by plasma, ultraviolet, and other surface-grafting techniques.

Surface grafting by ultraviolet, plasma, or other means can also be used to improve blood and tissue compatibility with polymers, including films. In general, hydrophilic monomers such as acrylamide and hydroxy methyl meth-

TABLE II

Reverse Osmosis Results of Plasma-Polymerized Polymer from Hydrophilic Monomers.

Monomer	Substrate	NaCl Concentration (%)	Applied Pressure (psi)	Salt Rejection (%)	Water Flux (gfd)
4-vinylpyridine	porous glass	1.2	1,200	96	.81*
	porous glass	1.2	1,200	96	.72*
	polysulfone	1.2	1,200	97	3.7
	polysulfone	1.2	1,200	98	7.0
	Millipore-VS	1.2	1,200	99	38.
	Millipore-VS	1.2	1,200	95	24.0
N-vinylpyrrolidone	polysulfone	1.2	1,200	91	10.6
4-picolin	polysulfone	1.2	1,200	98	6.4
4-ethyl pyridine	polysulfone	1.2	1,200	98	9.6
4-picolin	polysulfone	3.5	1,500	96	7.6
4-methylbenzylamine	polysulfone	3.5	1,500	96	2.2
n-butylamine	polysulfone	3.5	1,500	94	2.7
4-vinylpyridine	polysulfone	3.5	1,500	97	4.9
4-vinylpyridine	porous glass	3.5	1,500	96	0.80*
3-5 lutidine	polysulfone	3.5	1,500	99	12.

*Water flux of porous glass tube is approximately 1.0 gfd.

acrylate are used for forming a hydrophilic layer or as a basis for reacting with heparin or other compatible substrate materials. Surface grafting has also been used to improve the adhesion of polymer films to metal and other materials.

Impact-Resistant Grafted Films

The importance of graft copolymers in the successful production of high-impact polymer blends was stressed earlier in this paper. High-impact polystyrene and A–B–S polymers, both containing substantial proportions of the corresponding graft copolymer, are also produced on a large scale as films and sheets. These are used in vacuum forming and other molding methods for packaging and other applications.

Along somewhat similar lines, clear, transparent high acrylonitrile and methacrylonitrile polymers containing grafted nitrile rubber have been successfully produced [15]. The addition of the grafted rubber was shown to result in a sharp increase in impact strength; 10 percent grafted rubber gave a twentyfold increase in impact strength, while the films retained their clarity. It is anticipated that this technology for the production of high-barrier, high-transparency and tough films will reach commercial realization in the near future.

Acknowledgment

The author wishes to thank Dr. H. Yasuda for his help with the sections on mosaic membranes and plasma-induced surface grafting.

References

1. Stannett, V., "Some Recent Advances in Pure and Applied Aspects of Graft Co-polymers," *J. Macromol. Sci., Pt. A. Chem.,* A4, no. 5 (1970), 1177.

2. Rieke, J.K. and Hart, G.M., "Properties of Polyethene–Acrylic Acid Graft Copolymers," *J. Polym. Sci., Part C,* no. 1 (1963), 117.

3. Rieke, J.K., Hart, G.M., and Saunders, F.L., "Graft Copolymers of Polyethylene and Acrylic Acid II," *J. Polym. Sci., Part C,* no. 4 (1964), 589.

4. Wellons, J.D., Williams, J.L., and Stannett, V., "Preparation and Characterization of Some Cellulose Graft Copolymers. Part IV. Some Properties of Isolated Cellulose Acetate–Styrene Graft Copolymers," *J. Polym. Sci., Part A-1,* 3 (1967), 1341.

5. Bradford, E.B. and Vanzo, E., "Ordered Structures of Styrene–Butadiene Block Co-polymers in the Solid State," *J. Polym. Sci., Part A-1,* 6 (1968), 1661.

6. LeGrand, D.G., "Mechanical and Optical Studies of Poly(Dimethylsiloxane) Bisphenol-A Polycarbonate Copolymers," *J. Polym. Sci., Part B,* 7 (1969), 579.

7. Myers, A.W., Rogers, C.E., Stannett, V., Szwarc, M., Patterson, G.S., Hoffman, A.S., and Merrill, E.W., "The Permeability of Some Graft Copolymers of Polyethylene to Gases and Vapors," *J. Appl. Polym. Sci.,* 4 (1960), 159.

8. Huang, R.Y.M. and Kanitz, P.J., "Permeation of Gases Through Modified Polymer Films. I. Polyethylene–Styrene Graft Copolymers," *J. Appl. Polym. Sci.,* 13 (1969), 669.

9. Huang, R.Y.M. and Kanitz, P.J., "The Permeation of Gases Through Modified Polymer Films. IV. Gas Permeability and Separation Characteristics of Graft Copolymers of Polyethylene and Teflon FEP Films," *J. Appl. Polym. Sci.,* 15 (1971), 67.

10. Williams, J.L. and Stannett, V., "The Diffusion of Gases and Water Vapor Through Grafted Polyoxymethylene," *J. Appl. Polym. Sci.,* 14 (1970), 1949.

11. Hopfenberg, H.B., Kimura, F., Rigney, P.T., and Stannett, V., "Preparation and Proper-ties of Grafted Membranes for Desalination. Part I," *J. Polym. Sci., Part C.,* no. 28 (1969), 243.

12. Hopfenberg, H.B., Stannett, V., Kimura, F., and Rigney, P.T., "Novel Membranes Prepared by Radiation Grafting of Styrene to Cellulose Acetate," *Appl. Polym. Symp.,* no. 13 (1970), 139.

13. Bentvelzen, J., Kimura-Yeh, F., Hopfenberg, H.B., and Stannett, V., "Modification of Cellulose Acetate by Reverse Osmosis Membranes by Radiation Grafting," *J. Appl. Polym. Sci.,* 17 (1973), 809.

14. Boy, R.E., Schulken, R.M., and Tamblyn, J.W., "Crystallinity in Secondary Cellulose Esters," *J. Appl. Polym. Sci.,* 11 (1967), 2543.

15. Hughes, E.C., Idol, J., Duke, J.T., and Wick, L.M., "Transparent Barrier Resins with High Nitrile Content," *J. Appl. Polym. Sci.,* 13 (1969), 2567.

14. Special Properties of Block and Graft Copolymers and Applications in Fiber Form

JETT C. ARTHUR, JR.

United States Department of Agriculture
New Orleans, Louisiana

ABSTRACT

Graft and block copolymers of textile fibers have been prepared by both free-radical and ionic-initiated copolymerization reactions with vinyl monomers and combinations of vinyl monomers. The effects of block and graft copolymerizations on macromolecular, morphological, and mechanical properties of textile fibers are discussed. Changes in macromolecular properties of fibers lead to changes in second-order transition temperatures and elastic recovery properties. Also, these changes may result in changes in surface properties of fibers, so that fibrous copolymers have soil and water repellency and soil-release properties. Changes in morphological properties of fibers lead to changes in macrostructure that are reflected in mechanical properties of the fibrous copolymers, such as abrasion resistance and elastic recovery. Specific experimental applications of block and graft copolymerization to cotton, the world's most-used textile, are outlined.

The three possible applications of block and graft copolymers are in films, fibers, and plastics. This discussion is concerned with special applications in fibers. For the purposes of this discussion, fiber is defined as a relatively flexible, macroscopically homogeneous body having a high ratio of length to thickness and a small cross-section. The properties of fibers are related to their organo-chemical structure, macromolecular structure, and supermolecular, or morphological, organization. Most useful textile fibers exhibit a degree of molecular orientation, that is, they are semicrystalline.

Until about 1940, practically all textile fibers were based on naturally occurring polymers such as cellulose and protein. In 1969 the world consumption of fibers was about 43 billion pounds, which included almost 33 billion pounds of cellulosic fibers. In 1970 the consumption of fibers in the United States was about 9.6 billion pounds, which included: 3.8 (cotton), 1.4 (regenerated cellulosics), 4.1 (man-made), and 0.25 (wool) billion pounds [1].

Many textile products, particularly those derived from naturally occurring polymers, are fiber–polymer composites. Usually, solutions containing polymers or polymer-forming reagents are padded on textile fabrics. Then the treated fabrics are dried and heated briefly at elevated temperatures to form and/or to "cure" the polymer within the fabric structure [2].

Man-made fibers, such as acrylic and modacrylic fibers, are probably random block polymers that contain 35–85 percent acrylonitrile and 15–65 percent vinyl chloride, vinylidene chloride, vinyl acetate, styrene, or acrylamide. An example of a fibrous block copolymer used commercially in specialty products is an elastic fiber formed by cross-linking a soft-segmented prepolymer. Aromatic diisocyanate is reacted with a macroglycol to yield a fibrous soft-segmented prepolymer. Cross-linking of the fibrous prepolymer with glycol yields a fibrous hard-segmented polyurethane that has elastic properties similar to those of rubber. When the prepolymer is cross-linked with a diamine, a fibrous hard-segmented polyurea, which has elastic properties, is also obtained [2].

Scope

The preparation of graft copolymers and graft block copolymers of cotton fibers is demonstrated by free-radical-initiated processes that rely on liquid–solid reactions. Emphasis is on problems of modifying the properties of an existing fibrous polymer while retaining its natural properties. Usually, if a different property is desired in man-made fibers, the organochemical and/or macromolecular structures of the fibrous polymer are altered during the preparation of the fibers.

Definition of Terms

In the context of this discussion, a "graft copolymer" is a combination of cellulose and a polymer that is difficult to separate by solvent extraction without first degrading the cellulose, usually by hydrolytic and/or oxidative depolymerization; "graft block copolymer" is a combination of cellulose and a block polymer that is difficult to separate by solvent extraction.

Cotton Fiber

The chemical structure of cotton consists of units of cellobiose with the empirical formula $(C_{12}H_{20}O_{10})_n$. The value of n for cotton cellulose ranges as high as 2,000 to 2,500. Cotton cellulose is polymolecular and its apparent molecular weight, determined by different methods, may have different values. Viscosity-average molecular weights for natural cotton cellulose are about 700,000. Cotton cellulose is polymorphic; that is, it crystallizes into different forms. Cellulose I is the most common form and consists of about 70-80 percent highly ordered or crystalline regions, and 20-30 percent much less ordered or amorphous regions. The basic monoclinic unit cell of natural cellulose I contains two cellobiose units and has axes a = 8.35 Å, b = 10.3 Å, c = 7.9 Å, and β = 84°. The other important form of cotton is cellulose II, which is obtained when a swelling agent, such as a solution of sodium hydroxide, is added to and then removed from cotton. The unit cell of cellulose II also contains two cellobiose units and has axes a = 8.14 Å, b = 10.3 Å, c = 9.14 Å, and β = 62°. The highly ordered regions comprise about 60 percent of cellulose II. Typical X-ray diffractograms of cotton cellulose I and II are shown in Figure 1. Incidentally, celluloses contained in alkaline-processed wood pulps and regenerated cellulosic fibers usually have cellulose II forms. The density of cotton cellulose I ranges from about 1.50 to 1.60 g/cc; the tensile strength of cotton ranges from about 70,000 to more than 100,000 psi. Common solvents for cellulose are cuprammonium hydroxide and cupriethylenediamine [3].

The morphological organization of cotton fiber consists of four principal parts from its external surface: the cuticle, the primary wall, the secondary wall, and the lumen. The cuticle is a thin layer of noncellulosic components on the surface of the fiber. The primary wall forms the cellulosic envelope, about 0.1-0.2μ thick, for the protoplasmic solution from which glucose is polymerized to yield cellulose. Most of the cellulose of the fiber is found in the secondary wall, about 1-4μ thick, and is deposited in successive layers during the growth periods of polymerization of glucose. A cross-sectional view of a swollen fiber shows these layers of cellulose in Figure 2. Both primary and secondary walls are subdivided into fibrils. In the primary wall these fibrils spiral both clockwise and counterclockwise at an angle of about 70° without reversing along the length of

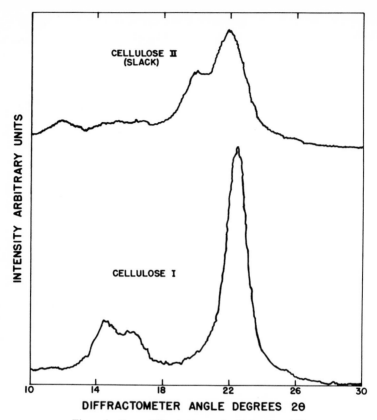

Figure 1. X-ray diffractograms of cellulose I and II.

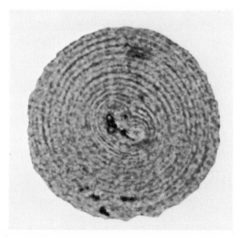

Figure 2. Cross-sectional view of swollen cotton fiber showing layers of cellulose.

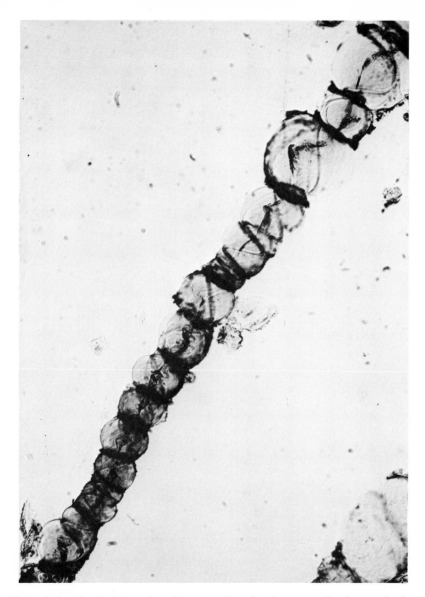

Figure 3. Longitudinal view of swollen cotton fiber showing supermolecular organization.

the fiber. In the secondary wall the fibrils spiral at angles of about 20°-45°, with reversing at frequent intervals along the length of the fiber. The lumen of the mature fiber, before the collapse of the growing fiber, is about 25-50 percent of the total cross-sectional area of the fiber. After collapse and dehydration

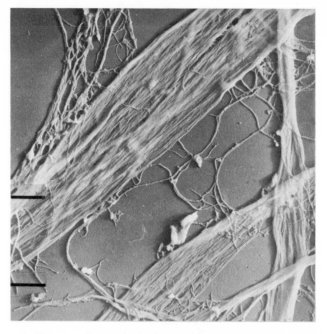

Figure 4. Electron micrograph of a fragment of cotton fiber showing fibrils.

of the fiber, the lumen is about 10 percent of this area. A longitudinal view of a swollen cotton fiber and an electron micrograph of a fragment of cotton fiber, which indicate this supermolecular organization, are shown in Figures 3 and 4, respectively [3].

Theory

Liquid–Solid Reactions

The formation of graft copolymers from fibrous polymers involves contacting a solution of vinyl monomer with fibrous polymer. Initiation of graft copolymerization of vinyl monomer with fibrous cellulose by ionic, charge-transfer, or free-radical processes has been reported [4,5]. In this discussion on the free-radical initiation of graft copolymerization reactions of vinyl monomers with cotton by ionizing radiation, both ultraviolet light and high-energy radiation are considered. In the case of cotton, two important factors in free-radical initiation of graft copolymerization reactions are the lifetime of the free-radical sites and the accessibility of the free-radical sites to the monomer [6,7].

When cotton cellulose is exposed to high-energy radiation, both short-lived and long-lived trapped cellulosic free radicals are formed. Solvent for the monomer can be selected so that a desired effect on the morphology of the cotton fiber can be obtained. In this way the rate and extent of graft copolymerization and the morphology of the product can be determined. When cotton cellulose is exposed to ultraviolet light, cellulosic free radicals are formed at which copolymerization reactions with vinyl monomers can be initiated. In the case of ceric ion, redox, or chemical oxidation processes, the lifetimes of the cellulosic free radicals are very short; monomer should therefore be present when the radicals are formed [6,7]. The reactivities of selected vinyl monomers for cellulosic free radicals, as compared with the reactivity of acrylonitrile when measured under the same experimental conditions, are shown in Table I [8].

TABLE I

Relative Molecular Reactivity of Vinyl Monomers for Trapped
Cellulosic Radicals in Irradiated Cotton* [8]

Vinyl Monomer	Relative Reactivity[†]
Acrylonitrile	1.00
Allyl methacrylate	0.17
Vinylpyrrolidone	0.24
1,3-butylene dimethacrylate	0.26
Lauryl methacrylate	0.29
Vinylidene chloride	0.45
Styrene	1.02
Butyl methacrylate	2.07
Glycidyl methacrylate	7.57
Methyl methacrylate	9.45

*Cotton preirradiated to a dosage of 1 Mrad.
[†]Experimental conditions (solvent, reaction time, and temperature) were the same for each vinyl monomer and acrylonitrile when relative molecular reactivity (moles monomer reacted/moles acrylonitrile reacted) was determined.

Living Polymer Radical

Electron spin resonance evidence for a living polymer radical formed during graft copolymerization of vinyl monomer with preirradiated cotton cellulose has been recorded [9]. A typical spectrum, generated by a trapped cellulosic radical in γ-irradiated, dried cotton cellulose I, is shown in Figure 5a. A sample of the irradiated cellulose was reacted with a deaerated aqueous solution of methacrylic acid (30 vol-%) for 3 min at 25°C. The unreacted monomer was then removed

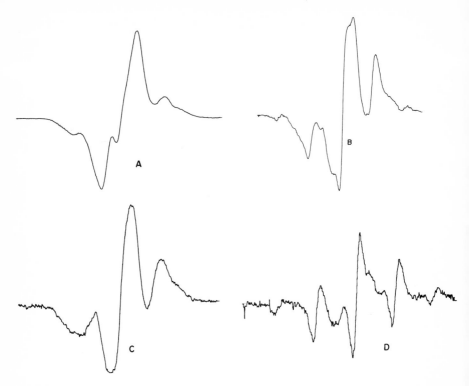

Figure 5. Electron spin resonance spectra of free radicals in copolymerization of meth-acrylic acid with irradiated cellulose (magnetic field sweep 100 gauss, left to right; center of spectrum *g*-value about free spin value): (a) irradiated, predried cellulose I; (b) cellulose–poly(methacrylic acid) copolymer; (c) product of (b) reacted with oxygen; (d) spectrum (b) minus spectrum (c).

from the cellulose graft copolymer product by extracting with deaerated water, followed by exchanging water with deaerated acetone and drying in a nitrogen atmosphere at 25°C. The ESR spectrum generated by this product is shown in Figure 5b. The graft copolymerization reaction could be reinitiated by contact-ing the product with solutions of the same or different monomers. When the product was extracted with aerated water, followed by exchanging the water with aerated acetone and drying in a nitrogen atmosphere at 25°C, it generated the ESR spectrum shown in Figure 5c. This is the typical ESR spectrum for trapped cellulose I radicals located within the highly ordered regions of the mac-rostructure. If this product is contacted with aqueous solutions of monomers, the copolymerization reaction is not reinitiated.

Using the time-averaging computer attachment, the ESR spectrum shown in Figure 5c was subtracted from that shown in Figure 5b to obtain the ESR spec-trum shown in Figure 5d. The ESR spectrum shown in Figure 5b is a composite of a spectrum generated by the living (or propagating) polymer radical and a spectrum generated by the trapped cellulose I radical located within the highly ordered regions of the macrostructure. Therefore, the ESR spectrum shown in Figure 5d should be generated by the living poly(methacrylic acid) radical. One set of five lines is located at about 0, ±23, and ±50 gauss; a weaker set of four lines is observed at about ±10 and ±43 gauss.

The living polymer radical could be generated as follows:

$$\text{cell} \cdot + n \quad \underset{\underset{H}{|}}{\overset{\overset{H}{|}}{C}} = \underset{}{\overset{\overset{CH_3}{|}}{C}} - COOH \longrightarrow \text{cell} - \left(\underset{\underset{H}{|}}{\overset{\overset{H}{|}}{C}} - \underset{\underset{COOH}{|}}{\overset{\overset{CH_3}{|}}{C}} \right)_{n-1} \underset{\underset{H}{|}}{\overset{\overset{H}{|}}{C}} - \underset{}{\overset{\overset{CH_3}{|}}{\dot{C}}} - COOH. \qquad (1)$$

There would be a number of growing poly(methacrylic acid) chains, originating at different cellulosic free-radical sites. At $25°C$ the methyl group should be freely rotating. For the polymer radical indicated in Equation (1) to generate a five-line ESR spectrum, the hydrogens of the methyl group and one of the methylene hydrogens would interact with the radical. In another conformation only the hydrogens of the methyl group apparently interacted with the radical to generate a four-line ESR spectrum. Based on the relative intensities of the four- and five-line spectra, the conformation that allowed one of the methylene hydrogens to interact with the radical was favored.

Model experiments were conducted in rigid glasses of methanol or acetone at near-liquid-nitrogen temperature. Propagating poly(methacrylic acid) radicals were generated by ferric-chloride-photosensitized initiation. The ESR spectrum of photolyzed acetone–methacrylic acid glass warmed from $-150°C$ to $-135°C$ for 2 min and then cooled to $-150°C$ is shown in Figure 6. A nine-line spectrum that consisted of a five-line spectrum with hyperfine splittings of about 0, ±22, and ±44 gauss and a four-line spectrum with hyperfine splittings of about ±11 and ±33 gauss was recorded.

When rigid glasses of methanol–methyl methacrylate, containing ferric chloride, were photolyzed for 2 min at $-170°C$, a triplet spectrum superimposed on another spectrum gave a five-line spectrum as shown in Figure 7a. The triplet spectrum was probably generated by the methanol radical. Interaction of this radical with methyl methacrylate, similar to the reaction proposed for cellulosic radical in Equation (1), gave a product that generated a five-line ESR spectrum. When the temperature of the photolyzed glass was increased to $-160°C$ for 2 min and then decreased to $-170°C$, an intensification of the five-line spectrum was recorded (Figure 7b). When the temperature of the glass was increased to $-150°C$ for 2 min and then decreased to $-170°C$, four additional lines appeared in the spectrum (Figure 7c). When the temperature of the glass was increased to $-140°C$ for 2 min and then decreased to $-170°C$, a nine-line spectrum was clearly generated (Figure 7d). The hyperfine splittings for the propagating

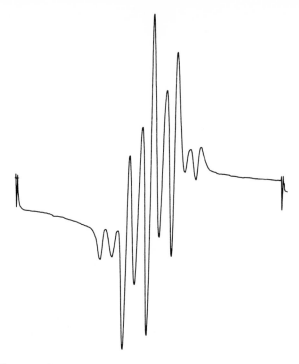

Figure 6. Electron spin resonance spectrum of propagating poly(methacrylic acid) radical generated in acetone glass at $-150°C$ by ferric-chloride-photosensitized initiation (spectral parameters same as in Figure 5).

methyl methacrylate radicals were the same as those reported for the methacrylic acid radicals.

It is suggested that one type of propagating vinyl monomer radical was formed. Initially, the conformation of the radical allowed the hydrogens of the methyl group and one of the methylene hydrogens to interact with the radical to generate a five-line spectrum. As the temperature of the rigid glass was increased, some of these radicals changed their conformations, so that only the hydrogens of the methyl group interacted with the radical to generate a four-line spectrum. The ESR spectrum of the living poly(methacrylic acid) radical shown in Figure 5d can be similarly interpreted.

Photoinitiated Polymer Radical

Direct photolysis of solutions of monomers did not yield any ESR-detectable radicals. However, when solutions of monomers were photolyzed in the presence of cotton cellulose, ESR spectra of propagating monomer radicals were re-

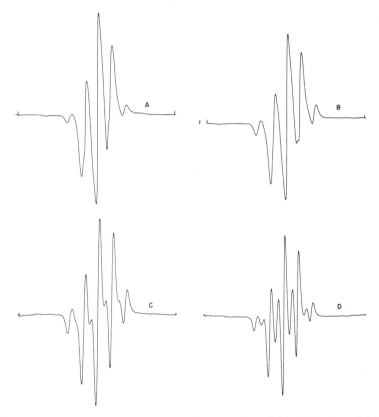

Figure 7. Electron spin resonance spectra of propagating poly(methyl methacrylate) radical generated in methanol glass at $-170°C$ by ferric-chloride-photosensitized initiation (spectral parameters same as in Figure 5): (a) methanol glass photolyzed for 2 min at $-170°C$; spectrum recorded at $-170°C$; (b) product of (a) warmed to $-160°C$ for 2 min, then spectrum recorded at $-170°C$; (c) product of (b) warmed to $-150°C$ for 2 min, then spectrum recorded at $-170°C$; (d) product of (c) warmed to $-140°C$ for 2 min, then spectrum recorded at $-170°C$.

corded as shown in Figure 8. The singlet ESR spectrum of dried cellulose that was photolyzed for 60 min at $40°C$ and recorded at $22°C$ is shown in Figure 8a. Cellulose was saturated with 0.5 M methacrylamide and dried. When this product was photolyzed under similar conditions, the five-line ESR spectrum shown in Figure 8b was obtained [10].

This radical would be generated as follows:

$$\text{cell} \cdot + n \quad \underset{\underset{H}{|}}{\overset{\overset{H}{|}}{C}} = \underset{\underset{CONH_2}{|}}{\overset{\overset{CH_3}{|}}{C}} - CONH_2 \longrightarrow \text{cell} - \left(-\underset{\underset{H}{|}}{\overset{\overset{H}{|}}{C}} - \underset{\underset{CONH_2}{|}}{\overset{\overset{CH_3}{|}}{C}} - \right)_{n-1} \underset{\underset{H}{|}}{\overset{\overset{H}{|}}{C}} - \underset{\overset{CH_3}{|}}{\overset{|}{\dot{C}}} - CONH_2. \quad (2)$$

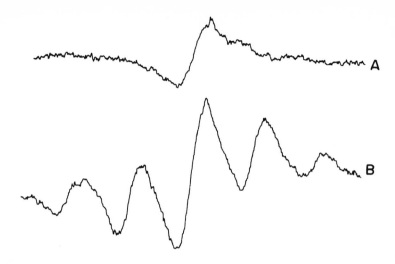

Figure 8. Electron spin resonance spectrum of propagating polymethacrylamide radical, generated in cellulosic matrix by photolysis at 40°C for 60 min, recorded at 22°C (spectral parameters same as in Figure 5): (a) electron spin resonance spectrum of photolyzed, dried cellulose I; (b) electron spin resonance spectrum of photolyzed, dried cellulose I that contained methacrylamide.

For the polymer radical indicated in Equation (2) to generate a five-line ESR spectrum, the hydrogens of the methyl group and one of the methylene hydrogens would interact with the radical. There was no evidence for an additional four lines in this spectrum as there was in the spectra shown in Figures 5 and 6 for propagating methacrylic acid and methyl methacrylate radicals. However, the hyperfine splittings of the five-line spectrum generated by the propagating methacrylamide radical were about the same as those of the similar five lines in the spectra generated by propagating methacrylic acid and methyl methacrylate radicals (see Figures 5 and 6).

In model experiments, the propagating methacrylamide radical was photolytically generated by ferric-chloride-sensitized initiation in methanol–methacrylamide rigid glasses at −160°C. Initially, the triplet spectrum of the methanol radical predominated, as shown in Figure 9a. When the temperature of the irradiated glass was increased to −140°C for 6 min and then decreased to −160°C, a symmetrical five-line spectrum was recorded, as shown in Figure 9b. When the temperature of the glass was further increased, only a five-line spectrum was recorded. For the case of methacrylamide radicals, the amide group apparently formed a structure that restricted rotation about the C_α-C_β bond, so that only the initial conformation, which generated a five-line spectrum, was allowed.

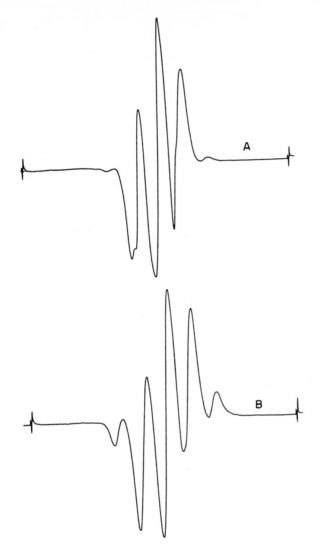

Figure 9. Electron spin resonance spectra of propagating polymethacrylamide radical generated in methanol glass at $-160°C$ by ferric-chloride-photosensitized initiation (spectral parameters same as in Figure 5): (a) methanol glass photolyzed for 3 min at $-160°C$; spectrum recorded at $-160°C$; (b) product of (a) warmed to $-140°C$ for 6 min; spectrum recorded at $-160°C$.

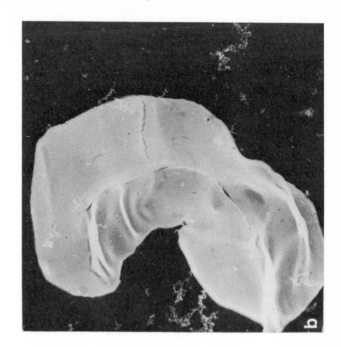

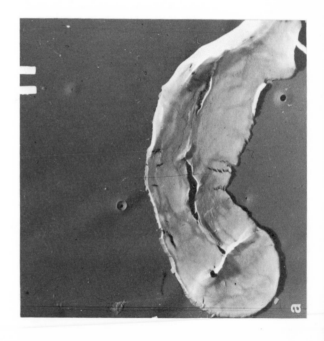

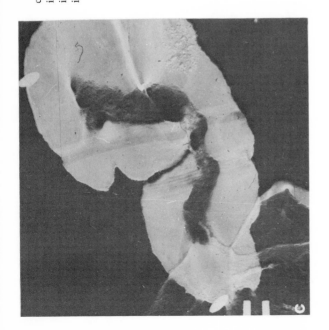

Figure 10. Electron micrographs of cross-section of fibrous cotton cellulose–polyacrylonitrile graft copolymers: (a) radiation-initiated copolymerization from N,N-dimethylformamide; (b) ceric-ion-initiated copolymerization from aqueous solution; (c) radiation-initiated copolymerization from saturated aqueous zinc chloride.

Grafted Block Copolymers

Grafted block copolymers of cellulose were prepared by irradiating dried cellulose to yield trapped cellulosic radicals and then immersing the irradiated cellulose in a solution of two or more monomers [8]. The following types of free-radical reactions could occur with preirradiated cellulose:

$$\text{cell·} \quad + X \longrightarrow \text{cell–}X· \tag{3}$$

$$\text{cell·} \quad + Y \longrightarrow \text{cell–}Y· \tag{4}$$

$$\text{cell–}Y· + X \longrightarrow \text{cell–}Y\text{–}X· \tag{5}$$

$$\text{cell–}X· + Y \longrightarrow \text{cell–}X\text{–}Y· \tag{6}$$

$$\text{cell–}Y· + Y \longrightarrow \text{cell–}Y\text{–}Y· \tag{7}$$

$$\text{cell–}X· + X \longrightarrow \text{cell–}X\text{–}X· \tag{8}$$

where cell· is cellulosic radical, and X and Y are monomers and may represent more than one monomer unit [8]. The relative molecular reactivity of the monomers for the cellulosic radical (see Table I) and the monomer reactivity ratios would determine the predominating reactions. For example, addition of styrene, 1,3-butylene dimethacrylate, or vinylpyrrolidone to acrylonitrile solutions in contact with irradiated cellulose increased the extent of graft copolymerization to a maximum value; addition of vinylidene chloride or allyl methacrylate to acrylonitrile did not greatly affect the extent of graft copolymerization; and addition of methyl or glycidyl methacrylate to acrylonitrile increased the extent of graft copolymerization.

Properties

Morphology

The morphology of cotton cellulose graft copolymers in the fiber form is dependent upon the method of free-radical initiation of the graft copolymerization reaction, the solvent used and its interaction with the cellulosic fibrous structure, and the monomer used [11]. The effects of method-of-initiation and solvent-used on the morphology of fibrous cotton cellulose–polyacrylonitrile copolymers are shown in Figures 10 and 11. The effects of solvents for the monomer that do not cause swelling of the collapsed natural fibrous macrostructure are shown in Figure 10. For example, the structure shown in Figure 10a was obtained by radiation-initiated grafting of acrylonitrile from N,N-dimethylformamide onto cotton cellulose. The structure shown in Figure 10b was obtained by ceric-ion-initiated grafting of acrylonitrile from aqueous solution onto cellulose.

The structure shown in Figure 10c was obtained by radiation-initiated grafting of acrylonitrile from saturated aqueous solution of zinc chloride (about 80 percent at 25°C).

The effects of solvents for monomers and of modifications of cellulose that cause swelling of the fibrous macrostructure are shown in Figure 11. For example, when cotton cellulose was cyanoethylated and then grafted with acrylonitrile by radiation initiation, the structure shown in Figure 11a was obtained. The structure shown in Figure 11b was obtained by first swelling the fiber in a solution of monomer (32 percent acrylonitrile in 80 percent aqueous zinc chloride at 25°C) and then grafting acrylonitrile onto cellulose by radiation initiation. The structure shown in Figure 11c was obtained by radiation-initiated grafting of acrylonitrile from an aqueous solution (15 percent acrylonitrile in 75 percent aqueous zinc chloride at 25°C).

The effects of monomer on the morphology of fibrous cellulose copolymers are shown in Figures 10–12. The structures shown in Figure 12 were obtained by first swelling fibers in methanolic solutions of monomers (30 percent monomer and 70 percent methanol) and then grafting monomer onto cellulose by radiation initiation. Layering or opening effects of the cellulosic structure were obtained for styrene (Figure 12a), vinyl acetate (Figure 12b), and hexyl methacrylate (Figure 12c).

Physical Properties

Elastic recovery properties. Typical elastic recovery properties of fibrous cellulose graft copolymers are shown in Table II. Radiation-initiated grafting of monomer onto cellulose was used in both cases. In the preparation of fibrous cellulose–polyacrylonitrile copolymer, the radiation dosage was low, so that the cellulose contained in the copolymer had a high molecular weight. In the preparation of fibrous cellulose–poly(ethyl acrylate), the radiation dosage was high, so that the cellulose contained in the copolymer had a low molecular weight. After elongation, both types of fibrous copolymers exhibited a high degree of elastic recovery. The fibrous cellulose–polyacrylonitrile copolymer had properties useful for textiles. The cellulose–poly(ethyl acrylate) copolymer had the characteristics of an elastomer [12].

Mechanical properties. Typical properties of cotton cellulose–polyacrylonitrile yarns grafted to different polymer add-ons are shown in Table III. The breaking strength of the yarn was not dependent upon polymer add-on and ranged from 3.7 to 4.4×10^3 g. However, the breaking stress and average stiffness of the yarn decreased with increased polymer add-on [12].

Fabric properties. Abrasion resistance. Cotton cellulose copolymer fabrics have improved abrasion resistance. Scanning electron micrographs of fabrics be-

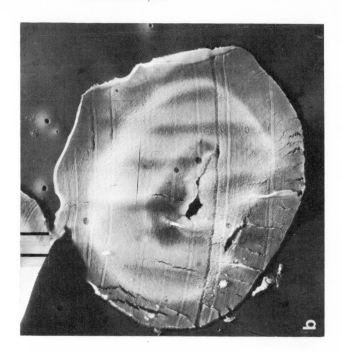

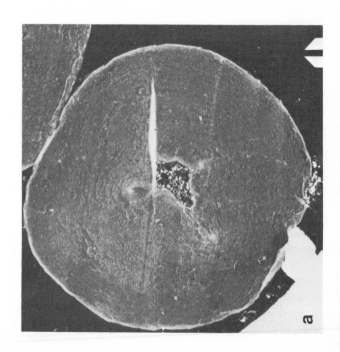

Figure 11. Electron micrographs of cross-sections of fibrous cotton cellulose–polyacrylonitrile graft copolymers: (a) radiation-initiated copolymerization with cyanoethylated cotton (D.S. 0.7) from saturated aqueous zinc chloride; (b) fiber preswollen in solution of monomer and then radiation-initiated copolymerization; (c) radiation-initiated copolymerization from solution of zinc chloride that caused swelling of the fiber.

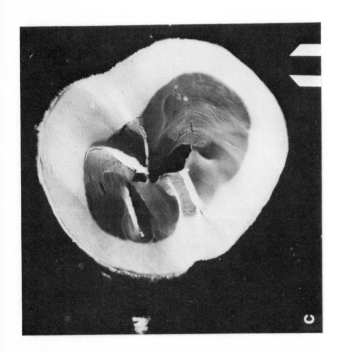

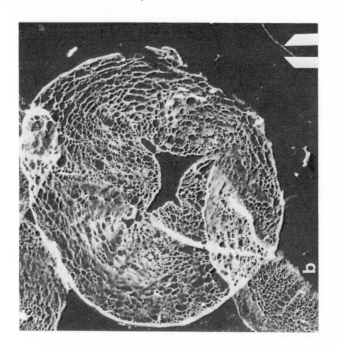

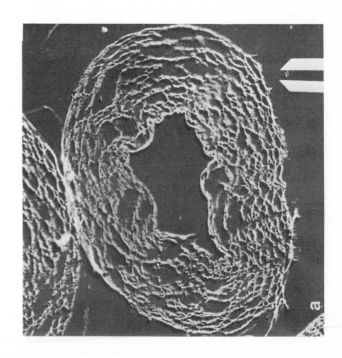

Figure 12. Electron micrographs of cross-sections of fibrous cotton cellulose graft copolymers, radiation-initiated copolymerization from methanol: (a) cellulose–polystyrene; (b) cellulose–poly(vinyl acetate); (c) cellulose–poly(hexyl methacrylate).

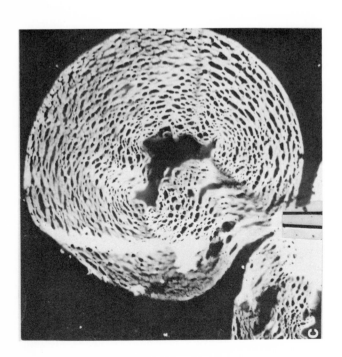

TABLE II

Elastic Recovery Properties of Cellulosic Fibers Containing
Graft Copolymers, Measured at 25°C [12]

Elongation (%)	Fraction of Elongation Recovered (%)	
	Cellulose–Polyacrylonitrile*	Cellulose–Polyethyl acrylate[†]
1.4	97	–
2.8	95	–
7	83	–
10	–	100
14	53	–
21	33	–
50	–	97
100	–	94
300	–	90
500	–	88

*Cellulose–polyacrylonitrile, 38%; dosage, 1 Mrad; degree of polymerization of cellulose, 800.
[†]Cellulose–poly (ethyl acrylate), 300%; dosage, 50 Mrads; degree of polymerization of cellulose, 40.

fore and after abrasion are shown in Figures 13 and 14. A cotton fabric is shown in Figure 13a. After flex abrasion testing to 2,000 cycles, the fabric is ruptured, as shown in Figure 13b. A cotton cellulose-polyacrylonitrile-poly(butyl methacrylate) fabric is shown in Figure 14a. After abrasion testing to 24,000 cycles, the fabric exhibits damage, as shown in Figure 14b. However, the damaged co-

TABLE III

Properties of Cotton Yarns Containing Cellulose Graft Polymers* [12]

Polymer Add-on (%)	Breaking Strength (10^{-3} g)	Breaking Stress (g/tex)	Average Stiffness (g/tex)
0[†]	4.4	18	116
0[‡]	4.2	17	127
11	3.9	14	108
24	3.7	12	81
57	4.4	10	46
77	3.8	8	35
105	4.1	7	25

*Yarn irradiated to a dosage of 1 Mrad, and then graft copolymer formed by reaction of irradiated yarn with acrylonitrile (32%) dissolved in zinc chloride (80%) solution at 25°C for time required to yield product with indicated polymer add-on. Measured at 25°C.
[†]Control yarn, 253 tex.
[‡]Irradiated control yarn.

TABLE IV

Soil-Release Properties of Cotton Cellulose I–Poly (methacrylic acid) Copolymer Fabrics* [12]

Polymer Add-on (%)	Oily Soil		Aqueous Soil	
	Soiling (%)	Soil Removed (%)	Soiling (%)	Soil Removed (%)
Control Fabrics				
0	73	41	51	41
8	72	50	52	49
10	69	50	48	43
19	71	66	58	50
Cross-linked Fabrics				
0	75	13	65	35
7	73	54	70	62
13	73	56	44	57
22	73	69	43	52

*Cross-linked with dimethyloldihydroxyethylene urea; initial cotton fabric weight, 108 gm^{-2}.

TABLE V

Durable-Press Properties of Cross-linked Cotton Fabrics Containing Cellulose Graft Copolymers* [12]

Grafted Monomer	Polymer Add-on (%)	Wash–Wear Rating After Wash–Dry Cycle 30	Flat Abrasion (fill) (cycles)	Wrinkle Recovery Angle (W + F) (degree) (conditioned)
Twill Fabric				
–	0	4.0	290	263
Acrylonitrile	8	4.5	520	252
Methyl methacrylate	14	4.5	400	286
Butyl methacrylate	6	4.6	470	278
Lauryl methacrylate	4	4.2	400	279
Print Cloth Fabric				
–	0	4.4	110	274
Acrylonitrile	9	4.4	90	293
Methyl methacrylate	13	3.6	90	303
Butyl methacrylate	4	4.1	120	311

*Cross-linked with dimethyloldihydroxyethylene urea; initial twill fabric weight, 254 gm^{-2}; initial print cloth fabric weight, 108 gm^{-2}.

polymer fabric does not exhibit rupture, as shown for the abraded, untreated fabric [12].

Soil release. Modification of the properties of cotton fabrics by cross-linking with polymer-forming finishes usually decreases the ease of soil removal from the textiles. The preparation of cotton cellulose copolymers with vinyl monomers, which have functional groups that are hydrophilic in nature, yields a tex-

Figure 13. Scanning electron micrographs of abraded cotton fabric: (a) control fabric; (b) fabric after abrasion.

tile product with improved soil-release properties, as shown in Table IV. Cross-linked, untreated fabrics exhibit a decrease in oily-soil- and aqueous-soil-removal properties, as compared with untreated control fabrics. Cross-linked fabrics that contained grafted poly(methacrylic acid) exhibited improved soil-removal properties, as compared with cross-linked control fabrics [12].

Durable press. The durable-press properties of cross-linked cotton fabrics that

contained cellulose graft copolymers are shown in Tables V and VI. Three different types of fabric construction were used: print cloth, twill, and twill woven from blends of natural and grafted fibers. In each case the cellulose graft copolymers were formed prior to cross-linking of the fabrics with dimethyloldihydroxyethylene urea. The wash–wear ratings of the cross-linked fabrics were satisfactory for textile uses. The cross-linked twill fabrics that contained cellulose graft

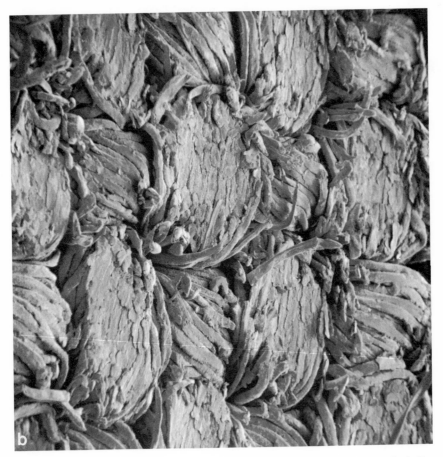

Figure 14. Scanning electron micrographs of abraded cotton cellulose–polyacrylonitrile–poly(butyl methacrylate) fabric: (a) control fabric; (b) fabric after abrasion.

copolymers had higher wash–wear ratings than cross-linked control fabrics. Also, in most cases the flat abrasion resistances and wrinkle recovery angles of the cross-linked twill fabrics that contained copolymer were equal to or greater than those of cross-linked control fabrics. The wrinkle recovery angles of cross-linked print cloth fabrics that contained copolymer were about $20°$–$35°$ greater than those of cross-linked control fabric [12].

TABLE VI

Durable-Press Properties of Cross-linked Twill Fabrics Woven
from Blends of Natural and Grafted Fibers [12]

Twill Fabric*	Average Polymer Content (%)	Wash–Wear Rating After Wash–Dry Cycle 30	Wrinkle Recovery Angle (W + F) (degree)	
			Conditioned	Wet
control†	0	4.1	264	247
A	6	4.7	275	259
B	6	4.6	248	254
C	11	4.3	270	248

*A: woven from blend of cotton–poly styrene (29%) fibers (1 part) + cotton (4 parts); B: woven from blend of cotton–poly styrene (12%) fibers (1 part) + cotton–poly styrene (18%) fibers (1 part) + cotton (3 parts); C: woven from blend of cotton–poly (methyl methacrylate) (53%) fibers (1 part) + cotton (4 parts).
†Control fabric weight, 305 gm^{-2}; fabrics cross-linked with diemthyloldihydroxyethylene urea.

Summary

Graft and block copolymers of cotton were prepared by free-radical-initiated copolymerization reactions of vinyl monomers with cellulose. The macromolecular, morphological, and textile properties of cotton (in fiber, yarn, and fabric forms) were modified. Generally, the mechanical, surface, and durable-press properties of cotton fabrics that contained copolymers were improved, as compared with those of control fabrics.

References

1. U.S. Department of Agriculture, *Agricultural Statistics,* Washington, D.C.: U.S. Government Printing Office (1971). (S/N 0100–1540)
2. Rubenfeld, L., "Fibers," *Encyclopedia of Polymer Science and Technology,* Vol. 6, 505.
3. Arthur, J.C., Jr., "Cotton," *Encyclopedia of Polymer Science and Technology,* Vol. 4, 244.
4. Arthur, J.C., Jr., "Graft Polymerization onto Polysaccharides," in *Advances in Macromolecular Chemistry,* Vol. 2, W.M. Pasika, ed., London: Academic Press (1970), 1.
5. "Proceedings of the Symposium on Graft Polymerization onto Cellulose," J.C. Arthur, Jr., ed., *J. Polym. Sci., Part C,* no. 37 (1972).
6. Arthur, J.C., Jr., "Cellulose Graft Copolymers," in *Addition and Condensation Polymerization Processes,* R.F. Gould, ed., Washington, D.C.: American Chemical Society (1969), 574.
7. Arthur, J.C., Jr. and Hinojosa, O., "Oxidative Reactions of Cellulose Initiated by Free Radicals," *J. Polym. Sci., Part C,* no. 36 (1971), 53.

8. Harris, J.A. and Arthur, J.C., Jr., "Radiation-Initiated Graft Copolymerization of Binary Monomer Mixtures Containing Acrylonitrile with Cotton Cellulose," *J. Appl. Polym. Sci.,* 14 (1970), 3113.

9. Hinojosa, O. and Arthur, J.C., Jr., "Propagating Radical in Copolymerization of Methacrylic Acid with γ-Irradiated Cellulose," *J. Polym. Sci., Part B,* 10 (1972), 161.

10. Harris, J.A., Hinojosa, O., and Arthur, J.C., Jr., "Conformational Effects of the ESR Spectra of Transient Vinyl Monomer Radicals," *Polym. Prepr., Amer. Chem. Soc., Div. Polym. Chem.,* 13, no. 1 (1972), 479.

11. Rollins, M.L., Cannizzaro, A.M., Blouin, F.A., and Arthur, J.C., Jr., "Location of Some Typical Vinyl Polymers within Radiation-Grafted Cotton Fibers: An Electron Microscopical Survey," *J. Appl. Polym. Sci.,* 12 (1968), 71.

12. Arthur, J.C., Jr., "Properties of Graft and Block Copolymers of Fibrous Cellulose," in *Multicomponent Polymer Systems,* Washington, D.C.: American Chemical Society (1971), 321.

Index